Communications
in Computer and Information Science 2398

Series Editors

Gang Li , *School of Information Technology, Deakin University, Burwood, VIC, Australia*
Joaquim Filipe , *Polytechnic Institute of Setúbal, Setúbal, Portugal*
Zhiwei Xu, *Chinese Academy of Sciences, Beijing, China*

Rationale

The CCIS series is devoted to the publication of proceedings of computer science conferences. Its aim is to efficiently disseminate original research results in informatics in printed and electronic form. While the focus is on publication of peer-reviewed full papers presenting mature work, inclusion of reviewed short papers reporting on work in progress is welcome, too. Besides globally relevant meetings with internationally representative program committees guaranteeing a strict peer-reviewing and paper selection process, conferences run by societies or of high regional or national relevance are also considered for publication.

Topics

The topical scope of CCIS spans the entire spectrum of informatics ranging from foundational topics in the theory of computing to information and communications science and technology and a broad variety of interdisciplinary application fields.

Information for Volume Editors and Authors

Publication in CCIS is free of charge. No royalties are paid, however, we offer registered conference participants temporary free access to the online version of the conference proceedings on SpringerLink (http://link.springer.com) by means of an http referrer from the conference website and/or a number of complimentary printed copies, as specified in the official acceptance email of the event.

CCIS proceedings can be published in time for distribution at conferences or as post-proceedings, and delivered in the form of printed books and/or electronically as USBs and/or e-content licenses for accessing proceedings at SpringerLink. Furthermore, CCIS proceedings are included in the CCIS electronic book series hosted in the SpringerLink digital library at http://link.springer.com/bookseries/7899. Conferences publishing in CCIS are allowed to use Online Conference Service (OCS) for managing the whole proceedings lifecycle (from submission and reviewing to preparing for publication) free of charge.

Publication process

The language of publication is exclusively English. Authors publishing in CCIS have to sign the Springer CCIS copyright transfer form, however, they are free to use their material published in CCIS for substantially changed, more elaborate subsequent publications elsewhere. For the preparation of the camera-ready papers/files, authors have to strictly adhere to the Springer CCIS Authors' Instructions and are strongly encouraged to use the CCIS LaTeX style files or templates.

Abstracting/Indexing

CCIS is abstracted/indexed in DBLP, Google Scholar, EI-Compendex, Mathematical Reviews, SCImago, Scopus. CCIS volumes are also submitted for the inclusion in ISI Proceedings.

How to start

To start the evaluation of your proposal for inclusion in the CCIS series, please send an e-mail to ccis@springer.com.

Abdulrahman Azab · Tomasz Malkiewicz
Editors

Nordic e-Infrastructure Tomorrow

6th Nordic e-Infrastructure Collaboration Conference
NeIC 2024, Tallinn, Estonia, May 27–29, 2024
Proceedings

Springer

Editors
Abdulrahman Azab
University of Oslo and Sigma2 AS
Oslo, Norway

Tomasz Malkiewicz
IT Center for Science Ltd.
Espoo, Finland

ISSN 1865-0929   ISSN 1865-0937 (electronic)
Communications in Computer and Information Science
ISBN 978-3-031-86239-7   ISBN 978-3-031-86240-3 (eBook)
https://doi.org/10.1007/978-3-031-86240-3

This work was supported by NordForsk.

This Springer imprint is published by the registered company Springer Nature Switzerland AG
The registered company address is: Gewerbestrasse 11, 6330 Cham, Switzerland

If disposing of this product, please recycle the paper.

Preface

The Nordic e-Infrastructure Collaboration 2024 conference (NeIC 2024), themed "Nordic e-Infrastructure Tomorrow," represented a pivotal opportunity to envision and shape the future of e-infrastructure in the Nordic region and beyond. NeIC, neic.no, is committed to fostering collaboration among Nordic and Baltic countries, ensuring that e-infrastructure services remain cutting-edge, interoperable, and accessible. This mission underpinned the conference, bringing together a diverse audience of researchers, practitioners, policymakers, and funders to share knowledge, discuss challenges, and explore innovative solutions.

At its core, NeIC 2024 reflected the evolving landscape of e-infrastructure, encompassing topics such as FAIR data stewardship, high-performance computing (HPC), sensitive data management, federated resource management, quantum computing, and AI-driven e-infrastructure solutions. Each of these areas underscores the shared goals of enhancing scientific discovery, improving data accessibility, and fostering cross-border collaboration. The conference tracks aimed to address specific challenges while highlighting opportunities for advancements that are both regionally relevant and globally impactful.

The scope of this year's conference papers spans a broad spectrum, from theoretical explorations to practical implementations. Key themes include the integration of HPC container technologies, the adoption of FAIR principles at both organizational and technological levels, and innovative approaches to secure data handling. These contributions not only advance the state of the art in e-infrastructure but also strengthen the collaborative fabric of the Nordic research community.

Conference Tracks and Themes

1. **FAIR Data Stewardship**: Exploring methods to ensure data is Findable, Accessible, Interoperable, and Reusable, this track highlighted best practices, standards, and challenges in data stewardship.
2. **High-Performance Computing and Storage**: Advancing the capabilities of HPC systems to meet the needs of large-scale, data-intensive research projects.
3. **Competence Building and Collaboration**: Fostering skills development and collaborative environments to maximize the utility of e-infrastructure resources.
4. **Sensitive Data Management**: Addressing the secure handling and processing of sensitive data within HPC ecosystems, ensuring compliance with privacy and confidentiality standards.
5. **Federated Resource Management**: Investigating strategies for efficient and secure cross-border resource sharing, including compute and storage solutions.
6. **AI and Machine Learning**: Showcasing innovations in e-infrastructure tailored to support AI and ML research.

7. **Quantum Computing**: Highlighting infrastructure solutions for quantum computing, including integration with existing HPC environments.
8. **Scientific Use Cases**: Presenting real-world applications that demonstrate the transformative potential of e-infrastructures.

The NeIC 2024 contributions in these proceedings underwent rigorous peer review, ensuring the highest scientific quality. Of 39 submissions, 14 full papers and 1 short paper were accepted. Each submission was single-blindly reviewed by three or more experts, reflecting the commitment of NeIC to maintain the integrity and relevance of its program.

We extend our heartfelt gratitude to the authors, reviewers, organizing committee, and sponsors whose efforts and expertise made NeIC 2024 a success. Together, we continue to build an e-infrastructure landscape that empowers research and innovation across the Nordic and Baltic regions.

December 2024

Abdulrahman Azab
Tomasz Malkiewicz

Organization

General Chair

Abdulrahman Azab   University of Oslo, Norway

Program Committee Chairs

Abdulrahman Azab   NeIC & University of Oslo, Norway
Tomasz Malkiewicz   NeIC & CSC - IT Center for Science, Finland

Program Committee

Anne Sofie Fink Kjeldgaard   Danish e-Infrastructure Consortium, Denmark
Ebba Þóra Hvannberg   University of Iceland, Iceland
Anders Sjöström   LUNARC, Sweden
Maria Francesca Iozzi   Sigma2 AS, Norway
Sebastian Von Alfthan   CSC - IT Center for Science, Finland
Kristina Lillemets   University of Tartu, Estonia
Oxana Smirnova   Lund University, Sweden
Bodil Aurstad   NordForsk, Norway
Josva Kleist   NORDUnet, Denmark

Organizing Committee

Ave Ploomipuu (Chair)   ETAIS, Estonia
Ülar Allas (Co-chair)   ETAIS, Estonia
Anders Sjöströms   Lund University, Sweden
Jarno Laitinen   CSC - IT Center for Science, Finland
Antti Valkonen   Aalto University, Finland
Lene Krøl Andersen   DTU Computerome, Denmark
Anne Fouilloux   Simula Research Laboratory, Norway
Russel Gene Wolff   University of Oslo, Norway
Vilma Häkkinen   NeIC, Finland
Mattias Wadenstein   HPC2N, Sweden

Review Board Members

Abdulrahman Azab	NeIC & University of Oslo, Norway
Tomasz Malkiewicz	NeIC & CSC - IT Center for Science, Finland
Ave Ploomipuu	ETAIS, Estonia
Ülar Allas	ETAIS, Estonia/UT HPC Center, Estonia
Anders Sjöström	LUNARC, Sweden
Alfio Lazzaro	HPE, Switzerland
Ebba Þóra Hvannberg	University of Iceland, Iceland
Barbara Krašovec	Jožef Stefan Institute, Slovenia
Dejan Lesjak	Jožef Stefan Institute, Slovenia
Abdelkader Berrouachedi	Université Paris 8, France
Adda Boualem	Ibn Khaldoun University in Tiaret, Algeria
Safia Deif	King Saud University|KKUH, Saudi Arabia
Hisham Kholidy	Northern Illinois University/SUNY Polytechnic Institute, USA
Maqsood Mahmud	Ulster University, UK
Andrew Treloar	Australian Research Data Commons, Australia
Chris De Loof	DARIAH, Belgium
Isabelle Perseil	Institut National de la Santé et de la Recherche Médicale, France
Matthew Dovey	JISC, UK
Ron Perrott	University of Oxford, UK
Sandra Gesing	University of Notre Dame, USA
Bengt Persson	NBIS, Sweden
Russel Wolff	UiO, Norway
Andreas Jeansen	Norway
Björn Grüning	University of Freiburg, Germany
Bob Killen	Lawrence Livermore National Laboratory, USA
Burak Yenier	UberCloud, USA
C. J. Newburn	NVIDIA, USA
Carlos E. Arango	Sylabs, USA
Giuseppa Muscianisi	CINECA, Italy
Martial Michel	Data Machines Corp, USA
Michael Jennings	Los Alamos National Lab, USA
Paolo Di Tommaso	Centre for Genomic Regulation/Seqera Labs, Spain
Parav Pandit	Mellanox, USA

Rosa M. Badia   Barcelona Supercomputing Center, Spain
Lucas Benedicic   CSCS, Switzerland
Carlos Fernandez   CESGA, Spain
Gilles Wiber   CEA, France
Diakhate Francois   CEA, France

Contents

A Novel Blockchain Approach for Continuously Authenticating UAVs to Achieve Zero Trust Principles in 5G-Enabled UAVs

Hisham A. Kholidy(✉)

Computer Science Department (Cybersecurity), Northern Illinois University, Dekalb, IL, USA
kholidy@niu.edu

Abstract. Drones can provide real-time data transmission, improved signal quality, and reduced interference through beamforming technology The integration of drones (also known as Unmanned Aerial Vehicles (UAVs)) with 5G connectivity technology is revolutionizing various industries by enabling advanced features and capabilities, especially the reliable and secure network coverage. Security is an essential aspect of any digital or communications system. In drone-powered communication systems, security poses an even greater challenge. This is mainly due to the automated nature of the system and remote wireless communication capabilities. For example, unlike terrestrial/ground base stations, if the flying mobile base station is compromised or attacked, it may lose its cellular connection with the ground station and destroy the UAV itself. One of the promising security solutions is Zero Trust Architecture (ZTA). The use of ZTA in drone/ Unmanned Aircraft Vehicle (UAV) security is still immature and has a lot of room for improvement. One particular problem is establishing continuous authentication mechanisms that can keep up with the changing operational dynamics of UAVs. Traditional security tactics frequently prove insufficient, either due to their lack of flexibility or failure to effectively handle the full range of possible dangers posed by UAVs.

This paper introduces a new approach to equip UAVs with a flying 5G base station to expand 5G coverage and connectivity in rural and limited coverage areas. To ensure secure connectivity in this ecosystem, we introduce a continuous Authentication-based Blockchain Approach that integrates blockchain's decentralization and transparency with a Zero Trust model we previously developed. The new architecture addresses one of the several obstacles that impede the integration of UAVs into the national airspace, which is real-time verification and airspace usage authentication. These criteria require the UAV to broadcast its identity to the 5G Base stations securely and trustworthy.

Keywords: Zero Trust · 5G · UAV · Drone · Authentication · Blockchain · Airborne

1 Introduction

5G-enabled UAVs can leverage the high-speed, low-latency, and reliable communication and control capabilities of 5G networks. This allows for remote control of drones, location tracking, video transfer via dedicated network slices, drone identification, and

© The Author(s) 2025
A. Azab and T. Malkiewicz (Eds.): NeIC 2024, CCIS 2398, pp. 1–16, 2025.
https://doi.org/10.1007/978-3-031-86240-3_1

route planning to avoid crowded areas. The integration of UAVs with 5G connectivity technology enhances the capabilities of drones in various applications [1] such as Aerial inspection and monitoring, film and entertainment, mapping and surveying, delivery and logistics, mission-critical operations, Disaster response and assessment, Precision agriculture and environmental monitoring, and Intelligence, surveillance, and reconnaissance (ISR). Additionally, drones operating on a 5G network can overcome challenges related to mobility, as they can be remotely controlled through a drone application even in areas with limited network coverage. The integration of 5G and drones in the context of mobility management allows for on-demand systems in specific regions, promoting accessibility to previously hard-to-reach areas [1].

The current state of development and implementation of UAV-enabled 5G connectivity technology shows significant progress. Several studies and industry demonstrations have showcased the benefits of using 5G networks to support various UAV applications, such as real-time data processing and offloading, secure data transfer, and seamless communication for autonomous drones. In the following, we highlight the added security features that 5G offers to the UAVs [2]:

- The air interface between the drone and the 5G base station is encrypted to ensure confidentiality and integrity of the communications over the wireless link.
- 5G employs a security edge protection proxy (SEPP) to verify and authorize traffic entering the network operator's internal network. This helps prevent unauthorized access and protect the core network.
- 5G uses a separation of duties and updated cryptographic key hierarchy to ensure that even if one sub-component is compromised, the rest of the 5G environment remains protected.
- 5G encrypts the 5G Subscription Permanent Identifier (SUPI) to conceal user identity and prevent tracking during the attachment procedure, addressing a key privacy weakness in prior cellular generations.
- The use of virtualization and network slicing in 5G enables security controls to be tailored for different data types and use cases, improving overall security and privacy.

However, 5G brings security concerns to the drone communications such as [2]:

- The reliance on legacy 4G/LTE protocols, which have known vulnerabilities, in non-standalone 5G deployments.
- The potential for downgrade attacks, where adversaries trick devices into using older, less secure cellular protocols.
- Risks introduced by cloud deployment and mobile edge computing, such as cloud misconfigurations and untrusted third-party code.

Overall, while 5G brings significant security improvements, there are still outstanding challenges that must be addressed to ensure the security of drone communications over 5G networks.

UAVs have the potential to revolutionize 5G deployment and usage by providing flexible, temporary, and emergency coverage in areas that are difficult to serve with traditional infrastructure by expanding coverage and strengthen connectivity in rural and low connectivity areas, enabling emergency response, and optimizing network infrastructure. Some key ways that drones can become 5G internet providers are [1]: (1) Airborne 5G Base Stations, (2) Drone Mesh Networks, (3) Connectivity for Emergency Response, (4) Flying WiFi Hotspots. Section 1.1 discusses various scenarios to equip Drones with 5G technology and highlights the several critical factors that must be addressed to successfully deploy drones as 5G internet providers which includes (1) coverage and reliability, (2) bandwidth and latency, and (3) regulatory compliance. To address these three challenges, this paper introduces two main contributions.

The first contribution of this paper leverages the Airborne 5G Base station approach [1] to equip the UAS with a flying 5G base station to expand 5G coverage and connectivity, improve bandwidth and latency, and enable the provision of temporary or permanent coverage in rural areas, disaster-stricken regions, and crowded events. The second contribution introduces a Blockchain based authentication approach of users, devices, and resources within the UAV ecosystem to comprehensively incorporate one of the major Zero Trust Principles namely continuous authentication and adaptive security measures across all dimensions of the UAV ecosystem. This comprehensive integration sets the stage for secure and resilient connectivity of UAVs into logistics and transportation networks to comply with reporting and control requirements. we will leverage our previously developed security techniques [3–5] to verify and authenticate the architecture entities (i.e., user, UAS, 5G base station, and communication channels).

The remainder of this paper is organized as follows. Section 1.1 discusses scenarios for equipping UAVs with 5G internet provision capability and related challenges. Section 1.2 discusses Zero Trust principles in the 5G networked drones. Section 2 introduces the two main contributions of the paper which include, the 5G networked drones' approach and the Blockchain-based authentication approach. Section 3 evaluates and discusses the proposed approaches. Finally, Sect. 4 draws some concluding remarks and outlines future work.

1.1 Equipping Drones with 5G Internet Provision Capability

There are two main paradigms for integrating UAVs into the cellular network, depending on the role that UAVs play [6]:

- Cellular-enabled UAV communication: UAVs with their own missions (e.g., cargo delivery, video surveillance) operate as new aerial users, which are served by the cellular GBSs.
- UAV-assisted cellular communication: UAVs operate as new aerial communication platforms (e.g., aerial base stations (BSs), relays) to serve the terrestrial users in the cellular network.

As shown in Fig. 1, the There are various scenarios to equip Drones with 5G technology to provide internet connectivity. Some key ways that drones can become 5G internet providers are [1]:

- **Airborne 5G Base Stations**: Drones can be used as airborne 5G base stations, enabling the provision of temporary or permanent coverage in rural areas, disaster-stricken regions, and crowded events. These flying cell towers can quickly and easily provide 5G coverage where traditional towers are not practical. Figure 1 depicts the hierarchical architecture of the Airborne 5G Base station for internet service provisioning.
- **Drone Mesh Networks**: Drones can be used to build a mesh network, acting as signal repeaters and boosters to expand the reach and functionality of 5G networks. The drones can dynamically adjust their positions to optimize coverage and capacity based on demand.
- **Connectivity for Emergency Response**: Drones with 5G capabilities can be rapidly deployed to restore communications during crises or natural disasters. They can improve coordination between first responders, victims, and support groups by providing temporary mobile cell towers.
- **Flying WiFi Hotspots**: In areas with large user densities, such as events or crowded public places, drones can be used to set up flying WiFi hotspots that leverage 5G to deliver high-speed internet connectivity.

To successfully deploy drones as 5G internet providers, several critical factors must be addressed [2]:

Coverage and Reliability: Ensuring reliable and ubiquitous 5G coverage is paramount for drone operations, as the loss of connectivity can result in the drone becoming lost or crashing. Strategies such as network bonding and private 5G networks can help mitigate coverage challenges.

Bandwidth and Latency: Drones involved in applications like inspection, surveillance, and search and rescue missions require sufficient bandwidth and low latency to enable the stable transmission of high-definition video and other data.

Regulatory Compliance: Strict airspace regulations pose challenges for drone operations, requiring secure and resilient connectivity to comply with reporting and control requirements.

1.2   The 5G-AKA protocol

5G-AKA (Authentication and Key Agreement) [7] is utilized within 5G networks to facilitate mutual authentication and establish shared security keys between the mobile device and the network. This protocol aims to secure communication by protecting various types of traffic, including NAS (Non-Access Stratum), RRC (Radio Resource Control), and User Plane traffic.

The authentication function of 5G-AKA ensures that both the mobile device (User Equipment or UE) and the network (home or serving network) are legitimate entities3. This two-sided authentication process prevents unauthorized access to services and mitigates risks associated with identity theft3.

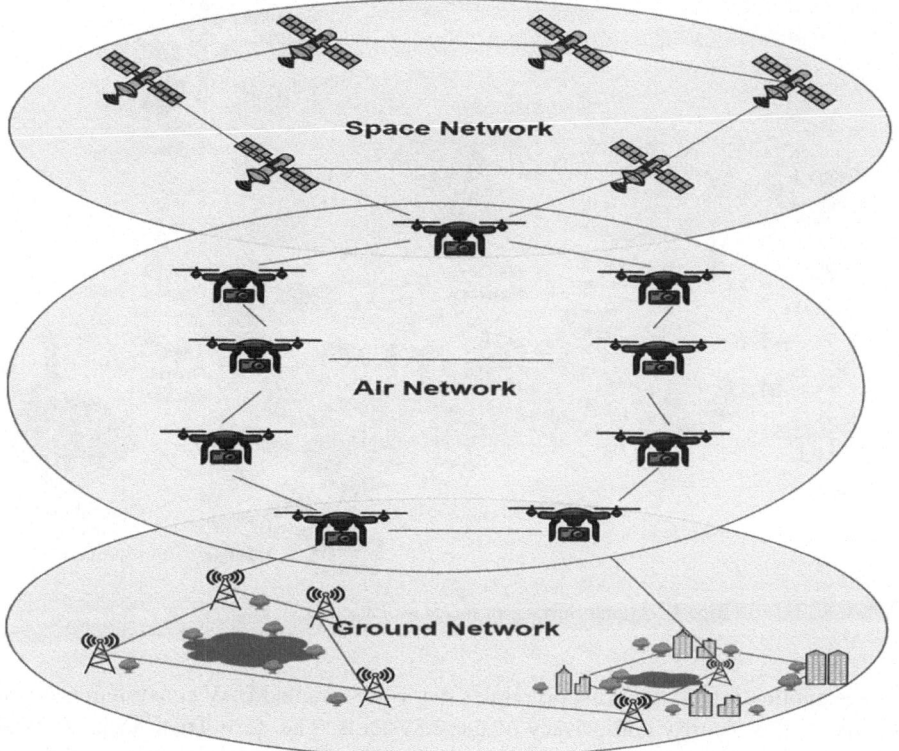

Fig. 1. The hierarchical architecture of the internet service provisioning Airborne 5G Base Stations

The key agreement aspect of 5G-AKA allows for the establishment of shared security keys crucial for cryptographic protection during communication sessions3. These keys help ensure that the data exchanged between the UE and the network remains confidential and intact, effectively preventing unauthorized interception.

1.3   Overview of Zero Trust Principles for UAV Systems

The unique vulnerabilities of drones and UAVs, such as Downlink threats that allow hackers to intercept data transmitted from the drone to the base station, posing risks to sensitive information and the zero-day vulnerabilities and potential hijacking threats that represent security gaps in drone communication networks and that allow attackers to gain unauthorized access and potentially hijack drones while airborne, make the application of zero trust principles crucial for this industry. By implementing a comprehensive zero trust framework, drone systems can enhance their overall security posture, mitigate the impact of cyber threats, and ensure the protection of sensitive data and assets (Fig. 2).

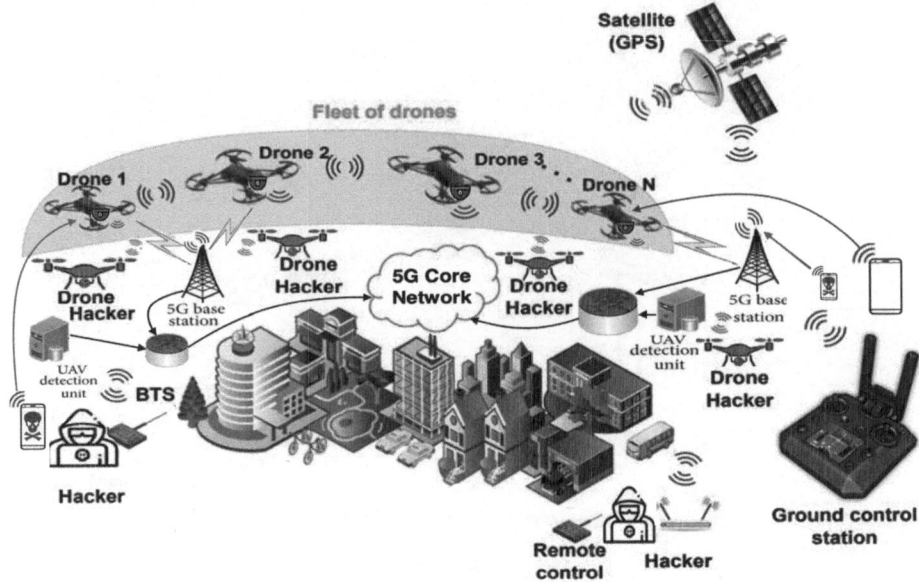

Fig. 2. Attack Surface in the 5G-enabled UAV ecosystem

The application of zero trust principles to the 5G-enabled UAV ecosystem is crucial to enhance the security and privacy of these systems. The Zero Trust Security model emphasizes the importance of continuous verification, limiting the potential impact of a breach, and automating the collection and response to contextual information. In the following, we briefly summarize the ZT main principles [4]:

1) Continuous authentication and verification in 5G-enabled UAV ecosystem

    In the context of the 5G-enabled UAV ecosystem, continuous verification involves rigorous authentication of all entities, including individuals and devices, that interact with the system. This eliminates the implicit trust granted based solely on network location or asset ownership. The verification process should consider factors such as user identity, device credentials, behavior patterns, and security posture to grant or maintain access to sensitive data and resources.

2) Limiting the Blast Radius

    A key principle of zero trust is to minimize the potential impact of a breach by limiting the "blast radius." This can be achieved in 5G-enabled UAV ecosystems through identity-based segmentation and the application of the least privilege principle. Specifically, access to sensitive information, such as parcel details and recipient data, should be restricted to only the authorized individuals or devices required to perform the delivery task.

3) Automated Context Collection and Response

    Automating the collection and analysis of contextual information is crucial in a zero-trust architecture for UAV delivery. This includes gathering data from various sources, such as user identities, device attributes, security events, and threat intelligence, to make accurate access decisions. By leveraging advanced analytics and machine learning, the system can rapidly detect and respond to anomalies or suspicious activities, enhancing the overall security posture.

4) Integrating Zero Trust in 5G-enabled UAV Ecosystem

Implementing zero trust principles in 5G-enabled UAV ecosystems involves adopting a comprehensive zero trust architecture (ZTA) that encompasses the entire system, from the ground control stations to the aerial vehicles and the associated cloud infrastructure. This approach ensures that all entities, including users, devices, and applications, are continuously verified and granted access based on their trustworthiness, regardless of their location or network connection.

One particular issue is establishing continuous authentication and verification mechanisms capable of keeping up with the evolving operational dynamics of drones. Traditional security tactics often prove inadequate, due to a lack of flexibility or inability to effectively manage all possible dangers posed by drones. Previous studies have investigated drone identification and classification using a variety of methods. For example, Kim et al. [8] and Choi and Oh [9] presented the use of convolutional neural networks (CNN) with micro-Doppler and Doppler images for drone classification. Similarly, other researchers have used deep belief networks to analyze surveillance images [10] and audio spectra [11]. However, these studies mainly focus on the detection component without fully integrating their approaches into a broader framework such as the ZTA.

Ekramulby et al. [12] used the Radio Frequency (RF) signals within a Deep Learning framework to identify drones to determine their permissions to access the UAVs network. In addition, the authors used the eXplainable Artificial Intelligence (XAI) tools such as SHapley Additive exPlanations (SHAP) and Local Interpretable Model-agnostic Explanations (LIME) to improve the model's transparency and interpretability. Dong et al. [13] introduced a seamless and continuous blockchain-based authentication of users, devices, and resources within the UAV ecosystem. This strategy ensures that access remains secure throughout all stages of engagement. The blockchain-based architecture guarantees data integrity and resilience, while advanced cryptographic techniques, including homomorphic encryption, ensure long-term security. Despite the last two studies adhering to the ZTA principles, they lack real-time interaction with the UAV network entities.

In this paper, we use the blockchain model to develop the proposed UAV authentication approach. Blockchain-powered drone communication faces several security issues such as (1) private key management, if the private key is stolen or lost, it can lead to unauthorized access and illegal transactions. (2) Malware and cyberattacks: Drones are susceptible to malware and cyberattacks, which can compromise the integrity and security of the blockchain network. (3) Network congestion: The large data volume that Drones can generate causes an overhead on the processing and storage on the blockchain network. (4) Scalability: adding more drones could exacerbate the Blockchain networks because it is not inherently scalable. (5) Regulatory Compliance: Blockchain-powered drone networks may be subject to regulatory compliance requirements, such as data security and privacy regulations.

2   Main Contributions and Methodology

2.1   The proposed connectivity provider architecture: Equipping UAV with 5G Internet Provision Capability

In this paper we use the Airborne 5G Base station approach where the UAVs can be used as airborne 5G base stations, enabling the provision of temporary or permanent coverage in rural areas, disaster-stricken regions, and crowded events. These flying cell towers can quickly and easily provide 5G coverage where traditional towers are not practical. To successfully deploy drones as 5G internet providers, several critical factors must be addressed [2]: (1) Coverage and Reliability. (2) Bandwidth and Latency, (3) Regulatory Compliance. The proposed connectivity provider infrastructure is designed to deliver seamless mobile connectivity through a robust system consisting of base stations, core networks, and intermediary routers/switches responsible for efficiently routing network packets over considerable distances beyond the base station. As illustrated in Fig. 3, this infrastructure maps out the user's pathway from the left-hand side to the cloud on the right. It's worth noting that the segment of infrastructure from the base station to the core network falls within the jurisdiction of the cellular service provider. However, expanding beyond the core network requires collaborative partnerships and agreements with other service providers to ensure continuous connectivity.

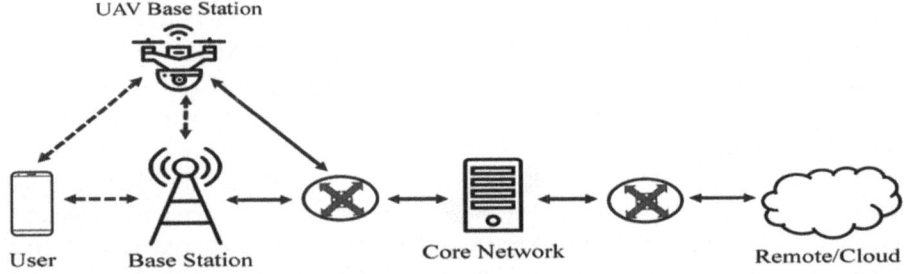

Fig. 3. The proposed connectivity provider architecture.

The base station serves as a pivotal link, wirelessly connecting to mobile users and acting as a conduit for wireless communications within telecom networks. It also serves as the intermediary between wireless and wired networking realms, transmitting RF signals to users while using wired connections to reach the internet and cloud. Responsibility for most digital processing tasks lies with the core network, including user registration, security setup, and access control. Data packets from users are processed within the core network before being sent to the internet. Figure 4 outlines proposed control communications among users, base stations, and the core network, with radio resource control (RRC) and a wireless channel essential for RF wireless communications. Securing a UAS flying base station is crucial due to its integral role in cellular service provider infrastructure, with a digital control process following initial configuration to ensure security and enable users to securely utilize RF and digital channels for data communications and networking applications. As Fig. 4 depicts, the digital control

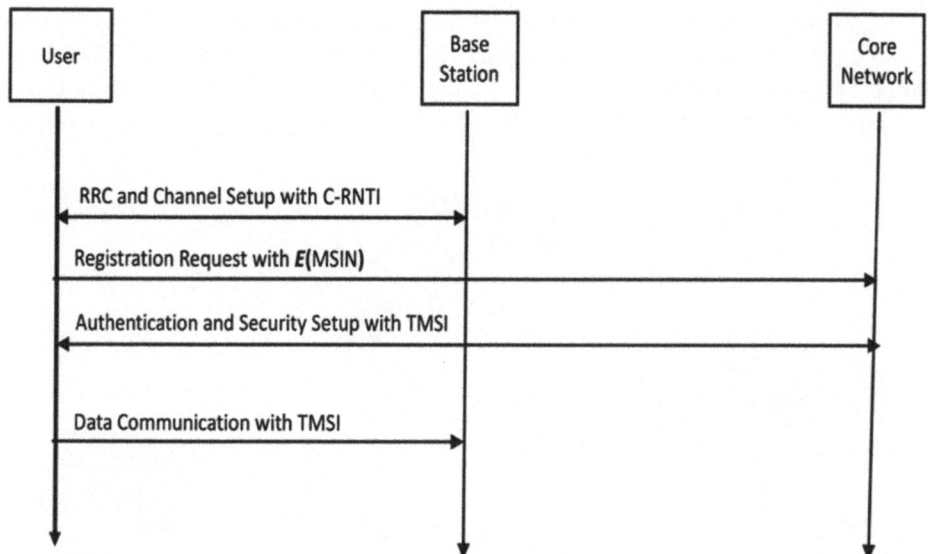

Fig. 4. The interactions between the connectivity provider architecture entities.

process includes the registration and authentication verification as well as the security setup between the user and the core network. From User and Subscriber Identity Module (Universal) registration (SIM or USIM) (which takes place before the user is enabled to receive connectivity services), the core network will retrieve the mobile subscriber identification number (MSIN) and a more temporary identifier from the temporary mobile subscriber identifier (TMSI) and share it with the user and base station. The MSIN is not transmitted in the clear but is processed using the encryption function E, thus exchanging E(MSIN). Users can then use the established RF and digital channels for network and data communications applications.

Important design ideas of the proposed system are briefly described as follows. Terrestrial UE can access the 5G core network via terrestrial 5G RAN. When terrestrial 5G RAN is operating normally, terrestrial UE can access the 5G core network by running an enhanced 5G-AKA protocol. When a particular area is not covered by the terrestrial 5G RAN or the terrestrial 5G RAN is located in the specific area fails, the terrestrial UE cannot access the 5G core network through the terrestrial 5G RAN. In this case, the remote terrestrial 5G RAN (a little further from the specific area) can deploy a drone to this specific area with low/no 5G coverage or the ground control station can command a drone to move to this area. The drone's communication range is wider than the regular UE, drones can access remote terrestrial RAN access (or forwarded by other radio access nodes such as satellite nodes) to receive network coverage. In this way, the drone helps the UE acquire much better network services. We also study a scenario where the drone's power is too low or Drones fail to provide the 5G coverage, other drones are quickly deployed by the remote terrestrial 5G RAN to replace the original drone. When the UE moves or the drone moves or is replaced, the UE and the target drone must perform a quick handover process to ensure continuous network service.

Following the 3GPP standards [14, 15] for User Equipment access to the 5G core network through the non-terrestrial network (UAVs, satellite, etc.), and the 3GPP standard [16] for the gNodeB onboard UAV, we developed the proposed 5G-enabled UAVs ecosystems architecture as depicted in Fig. 5.

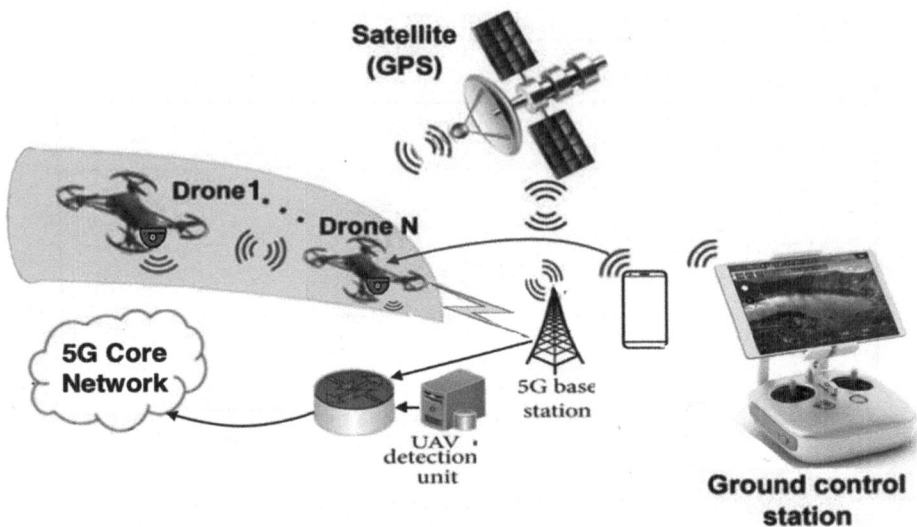

Fig. 5. The proposed 5G-enabled UAVs ecosystems architecture

As shown in Fig. 5, the drone is connected to the 5G core network via a 5G base station (terrestrial remote gNodeB). The UAV that helps the UE access the 5G core network is connected to a UAV detection unit that identifies the UAV, provides authentication and authorization services to the UE and the UAV and communicates with the authentication server function at the 5G core network. The 5G base station provides the communication interface between the drone/UE and the 5G core network. To reduce control latency and provide continuous network service to the UE, the Terrestrial RAN can control the start-up, shutdown, and replacement of the drone. Additionally, the UAV detection unit may store sensitive payload data (user authentication data, UAV operating status data, and location data). Terrestrial RAN can be deployed far away from the area to get 5G connectivity coverage. Drones can fly to a designated area and then hover over that area for a period of time. As a relay device, the UAV has the functions of base station and UE. First, the UAV accesses the network as a normal UE. The UAV then acts as a base station and provides network services to the UE. The maximum flight time of a drone is limited. Therefore, a drone cannot provide continuous network service, and replacement strategies must be in place. Additionally, because drones are easily damaged, the terrestrial RAN can monitor whether the drone is operating normally based on the frequency of regular heartbeat messages or inter-messages. When the RAN on the terrestrial detects that a drone has not sent a message for a certain period of time, other drones can be mobilized to replace the drone. UE represents a device that has base station access capability but does not have satellite access capability. The UE can access the 5G core network through the terrestrial RAN based on the existing 5GAKA protocol. In this architecture, terrestrial RAN and 5G core network entities are typically connected via wired links, which can establish a secure data channel through Datagram Transport Layer Security while existing data (DTLS). The others are connected through wireless links.

2.2 The Proposed Continuous Authentication Blockchain Approach

The proposed Continuous Authentication Blockchain approach is based on the current 5G-AKA protocol described in [17], see Fig. 6. The proposed approach authenticates the UAV to access the 5G core network and builds a secure channel between the UAV and the terrestrial RAN. Subsequently, the UE can execute the access authentication process to access the 5G core network and build a secure channel with the UAV.

When the UE uses the UAVs 5G connectivity capabilities provided by its flying 5G base station, the UAV provides access to the 5G core terrestrial network, then the remote terrestrial RAN initiates a scheduling notification to UAVs. Once notification is received by the UAVs, each UAV_j ready to provide network services shall perform the access authentication process, as shown in Fig. 7. The details are as follows.

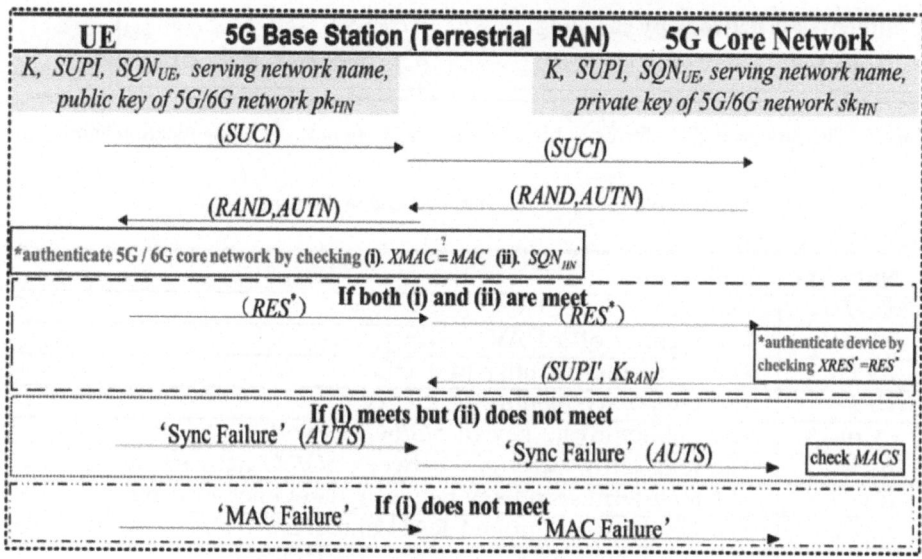

Fig. 6. The current 5G-AKA protocol

The first three steps are the same as that in the 5G-AKA protocol given above, except that the 5G core network needs to decide whether UAV_j is equipped with/ capable of offering 5G connectivity and informs the terrestrial RAN to authorize it.

Steps 4. If the terrestrial RAN authorizes the UAV_j, the terrestrial RAN generates one group key GK_{UAV} for all UAVs. After that, the terrestrial RAN generates a temporary identity TID_j, computes $C_j = EK_{RANj} (TID_j \|GK_{UAV})$, $HRES_j^* = KDF(KRAN_j, Cj \|RES_j^*$. The terrestrial RAN then transmits ($HRES_j^*$, C_j) to the UAV_j.

Steps 5. The UAV$_j$ computes the KRAN$_j$ and $XRES_j^*$ using the 5G-AKA protocol network by checking

$HXRES_j^* = HRES_j^*$. If that is trues, the UAVj calculates $TID_j' \parallel GK_{UAV}' = DKRANj$ (C$_j$) and SQN$_j$ = SQN$_j$ + 1, stores (TID$_j'$, KRANj, GK$_{UAV}'$) and confirm to the terrestrial RAN through an access confirmation message. Once this process is completed, the terrestrial RAN and the UAV$_j$ can be securely established (Table 1).

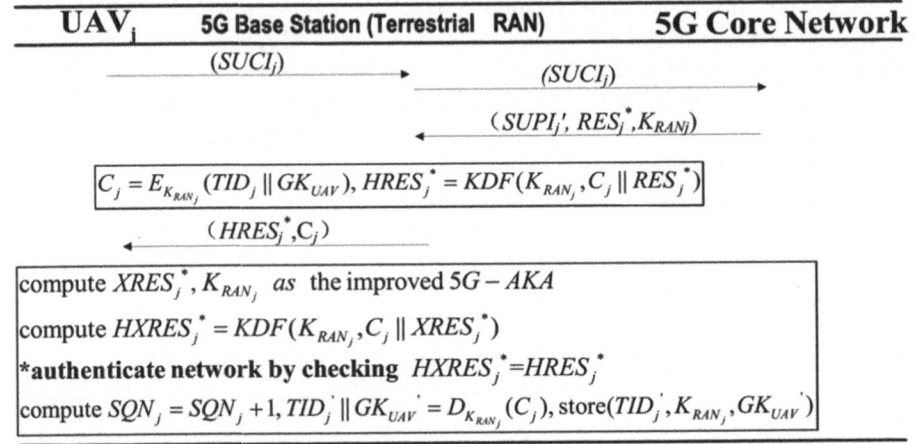

Fig. 7. The proposed UAV-Terrestrial Base station-5G core network authentication approach.

Table 1. Table of notations used in the paper

Notation	Definition
$SUPI_i$	The identity of a UE
$SUPI_j$	The identity of a UAV
TID_i	The temporary identity of a UE
TID_j	The temporary identity of a UAV
(A,pk)/sk	The public/private key of 5G/6G network
K_i	The long-term secret key between 5G/6G network and UE_i
K_j	The long-term secret key between 5G/6G network and UAV_j
K_{RAN_i}	The key between ground RAN and UE_i
K_{RAN_j}	The key between ground RAN and UAV_j
K_{UAV_i}	The key between UAV_j and UE_i
GK_{UAV}	The key shared among UAVs within a ground RAN
$Token$	A token used for handover
TS_T	The expiration time of $Token$
$H()$	Hash function $H : \{0, 1\}^* \rightarrow \{0, 1\}^{256}$
$E_k/D_k()$	Symmetric encryption/decryption function with key k

Once a secure session is established between the terrestrial RAN and the UAV$_j$, the payload will be stored in a blockchain and the smart contract will assess it with the newest regulations and information. If the smart contract approves payload data, an appropriate node in the 5G base station will be selected to generate an authentication credential for this payload. Figure 8 depicts the sequence of the blockchain authentication processes.

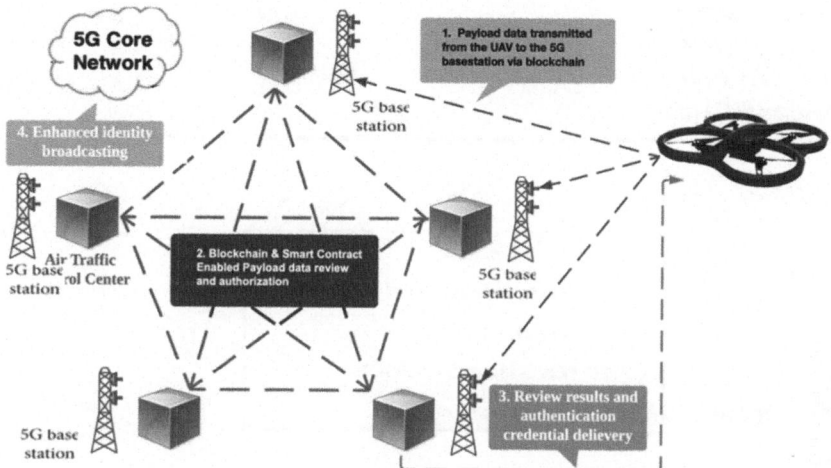

Fig. 8. The proposed Blockchain UAV authentication structure

3 Performance Evaluation

In this section, we evaluate the performance of the proposed approach and compare it with other related schemes with respect to the security properties and performance. Table 2 shows the security properties that the improved approach offers. Since there is no authentication scheme for the UAV relay to assist ground UEs in obtaining network services, we mainly compare the proposed scheme with the existing 5G-AKA protocol [7], and the existing satellite-assisted UE access authentication schemes, including Zhao et al.'scheme [18], Kumar and Garg scheme [19], Yang et al.'s scheme [20] and Guo and Du scheme [21]. To evaluate the performance of the improved scheme, we compare the communication overhead of relevant schemes that is caused by a large number of signaling messages and transmitted data during the authentication process that could potentially lead to network congestion, critical node failures, etc. Figure 9 depicts that our improved scheme has lower communication overhead compared to the existing schemes, except the Zaho et al. model which does not protect against traceability attacks and does not consider either quantum security or Perfect Forward Secrecy (PFS). The comparison shows that the added security properties that our improved protocol supports, including authentication, identity anonymity, data security, PFS, resistance to traceability attacks, ESL attacks, and protocol attacks, do not impact the performance much.

Table 2. Comparison of rSecurity Properties of relevant schemes.

Schemes	Mutual Authentication	Identity Anonymity	Traceability attack resistance	Data Security	PFS	ESL attack resistance	Protocol attacks resistance	Quantum Security
The existing 5G-AKA protocol [9]	√	√	X	√	X	√	√	X
Zhao et al.'s scheme [20]	√	√	X	√	X	√	√	X
Kumar et al.'s scheme [21]	√	√	√	√	√	X	√	√
Yang et al.'s scheme [22]	√	√	√	√	√	X	√	X
Guo et al.'s scheme [23]	√	√	X	√	√	√	√	√
Our improved 5G-AKA protocol	√	√	√	√	√	√	√	√

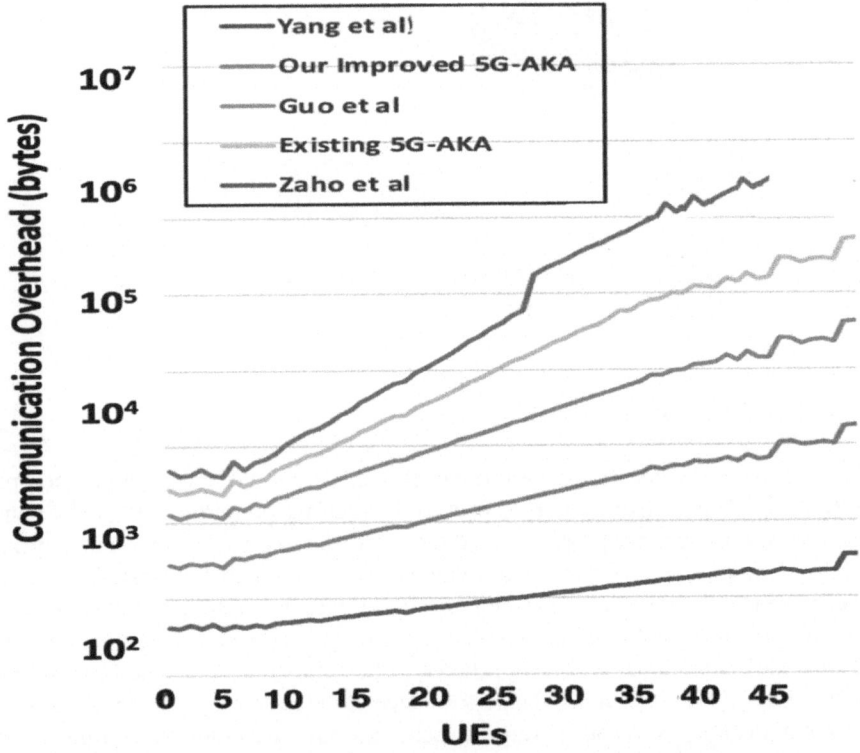

Fig. 9. Comparison of Communication Overhead of relevant authentication schemes.

4 Conclusion and Future Work

In this paper, we introduced a new approach to equip UAVs with a flying 5G base station to expand 5G coverage and connectivity in rural and limited coverage areas. To ensure secure connectivity in this ecosystem, we introduced a continuous Authentication-based Blockchain Approach that integrates blockchain's decentralization and transparency with a Zero Trust model we previously developed. The new architecture addressed one of the several obstacles that impede the integration of UAS into the national airspace, which is real-time verification and airspace usage authentication. Furthermore, we proposed a secure framework for the safe integration of the UAV using a blockchain-enabled framework for the authentication of UAV payload sensitive data and exchanged the authentication secrecy between the UAV, Terrestrial Base station, and 5G core network. For future work, we plan to augment this work to our zero-trust framework [24] to study the impact of the model on trust quantification.

Acknowledgment. This research was also supported in part by the Air Force Research Laboratory through the Information Directorate's Information Institute® contracts numbers FA8750-23-C-0517 and SA10032024050805" and was also supported by NIST through the IZUM company from funds received from the Korea Institute for Advancement of Technology (KIAT) (grant No., P0019809: "Building Enablers for Multi-Industry Sectors Collaborative Federated Testbeds, as a Foundation (Distributed Open Platform) for Cross-Industry End-to-End Services Innovation and Delivery Agility in the 5G & Beyond)". This research was also supported in part by the SUNY Polytechnic Institute 2024 Research Seed Grant Program and the 24 DOD appropriations project funded by the SUNY Polytechnic Institute.

References

1. Amponis, G., et al.: Drones in B5G/6G networks as flying base stations. Drones **6**, 39 (2022). https://doi.org/10.3390/drones6020039
2. Gilani, S.M., Anjum, A., Khan, A., Syed, M.H., Moqurrab, S.A., Srivastava, G.: A robust Internet of Drones security surveillance communication network based on IOTA. Internet Things **25**, 101066 (2024). ISSN 2542–6605. https://doi.org/10.1016/j.iot.2024.101066
3. Kholidy, H.A., et al.: Toward zero trust security in 5G open architecture network slices. In: MILCOM 2022 - 2022 IEEE Military Communications Conference (MILCOM), Rockville, MD, USA, pp. 577–582 (2022). https://doi.org/10.1109/MILCOM55135.2022.10017474
4. Kholidy, H.: Detecting impersonation attacks in cloud computing environments using a centric user profiling approach. Future Gener. Comput. Syst. **117**(17), 299–320 (2021), ISSN 0167–739X, https://doi.org/10.1016/j.future.2020.12.009. https://www.sciencedirect.com/science/article/pii/S0167739X20330715
5. Kholidy, H.A., Erradi, A., Abdelwahed, S., Azab, A.: A finite state hidden markov model for predicting multistage attacks in cloud systems. In: 2014 IEEE 12th International Conference on Dependable, Autonomic and Secure Computing, Dalian, China, pp. 14–19 (2014). https://doi.org/10.1109/DASC.2014.12
6. Zhang, S., Zeng, Y., Zhang, R.: Cellular-enabled UAV communication: a connectivity-constrained trajectory optimization perspective. IEEE Trans. Commun. **67**(3), 2580–2604 (2019). https://doi.org/10.1109/TCOMM.2018.2880468
7. https://telcomatraining.com/what-is-5g-aka-5g-authentication-and-key-agreement/
8. Kim, B.K., Kang, H.-S., Park, S.-O.: Drone classification using convolutional neural networks with merged doppler images. IEEE Geosci. Remote Sens. Lett. **14**(1), 38–42 (2016)
9. Choi, B., Oh, D.: Classification of drone type using deep convolutional neural networks based on micro-doppler simulation. In: 2018 International Symposium on Antennas and Propagation (ISAP), pp. 1–2. IEEE (2018)

10. Aker, C., Kalkan, S.: Using deep networks for drone detection. In: 2017 14th IEEE International Conference on Advanced Video and Signal Based Surveillance (AVSS), pp. 1–6 (2017)
11. Vemula, H.C.: Multiple drone detection and acoustic scene classification with deep learning. PhD thesis, Wright State University (2018)
12. Haque, E., Hasan, K., Ahmed, I., Alam, M.S., Islam, T.: Enhancing UAV security through zero trust architecture: an advanced deep learning and explainable AI analysis. In: IEEE ICNC 2024 (CNC Workshop) (2024)
13. Dong, C., et al.: Securing smart UAV delivery systems using zero trust principle-driven blockchain architecture. In: 2023 IEEE International Conference on Blockchain (Blockchain), pp. 315–322 (2023). https://doi.org/10.1109/Blockchain60715.2023.00056,
14. Technical specification group radio access network; study on new radio (NR) to support non-terrestrial networks; (Release 15). 3GPP, Sophia Antipolis, France, Rep. TR 38.811, V15.4.0 (2020)
15. Technical specification group radio access network; solutions for NR to support non-terrestrial networks (NTN); (Release 16). 3GPP, Sophia Antipolis, France, Rep. TR 38.821, V16.0.0 (2023)
16. Technical specification group services and system aspects; enhancement for unmanned aerial vehicles; stage 1; (Relese 17). 3GPP, Sophia Antipolis, France, Rep. TR22.829 V17.1.0 (2019)
17. Technical specification group services and system aspects; security architecture and procedures for 5G system; (Release 18). 3GPP, Sophia Antipolis, France, Rep. TR 33.501, V18.2.0 (2023)
18. Zhao, W., Zhang, A., Li, J., Wu, X., Liu, Y.: Analysis and design of an authentication protocol for space information network. In: Proceedings of IEEE Military Communications Conference, pp. 43–48 (2016)
19. Kumar, U., Garg, M.: Learning with error-based key agreement and authentication scheme for satellite communication. Int. J. Satell. Commun. 40(2), 83–95 (2022)
20. Yang, Q., Xue, K., Xu, J., Wang, J., Li, F., Yu, N.: AnFRA: anonymous and fast roaming authentication for space information network. IEEE Trans. Inf. Forensics Secur. 14, 486–497 (2019)
21. Guo, J., Du, Y.: A novel RLWE-based anonymous mutual authentication protocol for space information network. Secur. Commun. Netw. 2020, 1–12 (2020)

Quantization of Vision Transformer-Based Model for Real-Time EEG Classification

Rabindra Khadka[1,2(✉)], Poushali Sengupta[4], Pedro G. Lind[1,2,3], and Anis Yazidi[1,2]

[1] Oslo Metropolitan University (OsloMet), Oslo, Norway
rabindra@oslomet.no
[2] Nordic Centre for Sustainable and Trustworthy Artificial Intelligence Research (NordSTAR), Oslo, Norway
[3] Simula, Oslo, Norway
[4] University of Oslo, Oslo, Norway

Abstract. Electroencephalography (EEG) is a non-invasive and cost-effective tool for capturing brain signals. However, existing approaches based on deep learning models for classifying EEG signals are primarily trained on large EEG datasets and require extensive computational resources. This poses significant challenges for real-time processing, particularly in resource-constrained environments such as wearable devices and edge computing. Our work addresses these challenges by exploring quantization techniques to optimize a vision transformer model for real-time EEG classification. We apply post-training quantization to the proposed Vision Transformer (ViT) model, termed the Brain Signal Vision Transformer (BSVT), to reduce computational overhead while maintaining high performance. The BSVT model, trained on the ex3 HPC system, efficiently handles large datasets. We comprehensively analyze the trade-offs between model accuracy and computational efficiency. Specifically, we demonstrate the use of dynamic quantization on a ViT-based model for EEG classification, achieving a 2.49x reduction in model size and a 1.46x improvement in inference speed. These findings underscore the potential of quantization techniques and distributed computing to revolutionize the adoption of AI in healthcare.

Keywords: ViT · EEG · Quantization · Edge Computing

1 Introduction

Mobile edge computing has introduced a paradigm shift, particularly impacting applications requiring low latency and real-time decision-making [1]. One such application involves using EEG signals to diagnose and monitor neurological conditions. EEG captures dynamic brain activity through non-invasive electrodes placed on the scalp [2]. However, the real-time processing and classification of EEG data present significant challenges due to the high volume and complexity of these signals.

A. Azab and T. Malkiewicz (Eds.): NeIC 2024, CCIS 2398, pp. 17–27, 2025.
https://doi.org/10.1007/978-3-031-86240-3_2

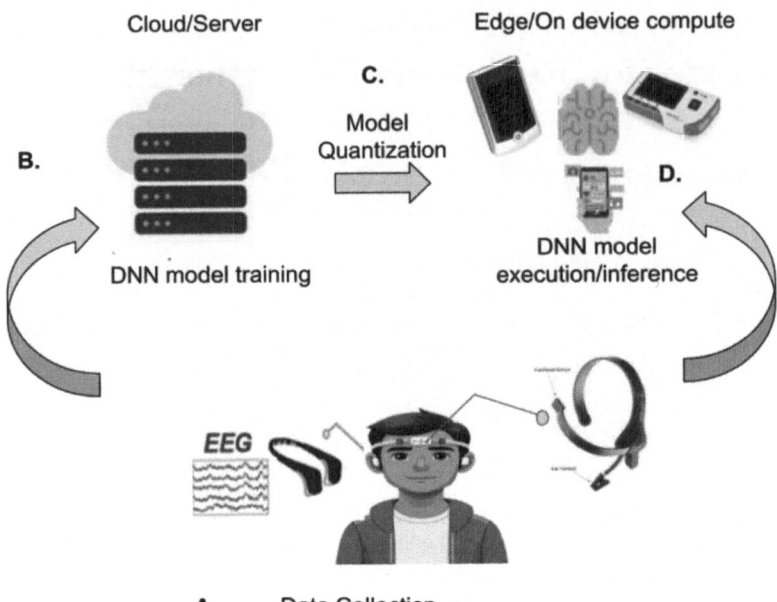

Fig. 1. The schematic setup for deploying deep neural networks for EEG analysis on edge devices. **A**. EEG data is collected from sensors. The data is transferred to the cloud/server. **B**. Data is cleaned, pre-processed, and stored on the cloud server. The deep learning model is trained and validated on the cloud. **C**. The selected model is quantized and deployed on different edge devices. **D**. The inference happens in real-time on the edge device given an incoming signal.

Current EEG signal analysis and classification methods often rely on centralized processing platforms [3]. This approach, however, introduces latency and dependence on reliable network connections, which are not always feasible in remote or mobile settings. Edge computing addresses these limitations by enabling local processing of EEG data on handheld devices such as smartphones, wearable technology, or portable medical equipment. This allows for faster response times and enhanced reliability in real-time EEG analysis, as illustrated by the setup schema in Fig. 1.

Despite its advantages, deploying complex machine learning models for EEG analysis on edge devices poses unique challenges. Edge devices typically have limited memory, computational power, and energy resources [4]. Consequently, efficient algorithms that account for these constraints are essential. Recent advancements in deep learning have established vision transformer (ViT) architectures as a popular choice for vision tasks. Studies have demonstrated that ViT-based models are effective for EEG classification [5–7]. The Brain Signal Vision Transformer (BSVT) [5] model is one such architecture adapted for training on EEG signals. However, vision transformers are resource-intensive, making them challenging to deploy on edge devices [8].

Quantization techniques are crucial in addressing this issue by reducing the size and computational requirements of deep learning (DL) models [9]. Quantization works by mapping high-precision floating-point values to lower-precision integers, enabling efficient deployment of ViT models on edge devices with minimal performance loss. In neural network quantization, the algorithms are broadly classified into two categories: (a) Post-Training Quantization (PTQ) and (b) Quantization-Aware Training (QAT). PTQ refers to a technique that converts a pre-trained FP32 network into a fixed-point network, while QAT incorporates quantization noise during the training process [10].

Despite the potential of these techniques, there is a lack of comprehensive studies exploring the application of vision transformers for EEG classification on edge devices using quantization methods. This gap underscores the need for research focused on adapting vision transformers for EEG signal processing while optimizing these models for deployment on edge devices without significant performance degradation. Addressing this challenge is vital for advancing real-time EEG analysis, with far-reaching implications for remote healthcare, neurological monitoring using mobile devices, and various other medical applications.

In summary, the key challenge is developing a vision transformer-based model for EEG classification that achieves a balance between accuracy and efficiency for deployment on edge devices. Our work addresses this gap by applying post-training quantization to vision transformers, leveraging its efficiency to compress models without the need for fine-tuning.

Our paper is structured as follows: Sect. 2 reviews previous studies related to our work. Section 3 details the methodology, including the model and quantization techniques, demonstrating the depth of our technical analysis. Section 4 presents the experimental results, highlighting implementation details and outcomes. Section 5 concludes the paper, summarizing our findings and their implications.

2 Related Work

Electroencephalograms (EEGs) capture the dynamical activity of the brain and have been widely adapted to understand disorders such as epilepsy and sleep disorders [11–13]. Owing to the success of the transformer-based model [14,15], researchers have applied the transformer-based model for various tasks. The work by [16] introduces the teacher-student strategy to train a transformer-based model showing competitive results without using the convolutional neural network. The Vision Transformer model has been widely adapted to tasks like semantic segmentation [17], object detection, and video analysis [18]. It has also been popular in applications related to EEG signal analysis. Some recent works adopt transformer-based models to classify EEG signals efficiently [5,19]. Despite their excellent performance, transformer-based models are data-hungry and require a lot of computation power. This poses significant constraints when deploying these models on edge devices. Hence, carefully considering power and memory requirements is necessary to integrate deep learning into EEG diagnosis tools for deployment on edge devices [20, 21]. AI-based medical tools can be made more portable and accessible across different levels of society across the globe by making them available on edge devices. The method proposed by [22] uses a multi-phase convolutional neural network embedded in edge devices to recognize emotions based on EEG signals, which has multiple applications for BCI. Similarly, the work by [23] implemented a framework on low-power IoT edge devices to analyze neonatal EEG signals for seizure detection, which empowers under-resourced communities across the globe. There are also efforts to develop real-time EEG analysis IoT devices for cognitive prediction using deep learning model [24]. A recent work by [25] proposed logic-in-headbands-based edge analytics (LiHEA) that can be integrated into head devices to reduce delay and bandwidth consumption.

Quantization is one of the most effective ways to deploy deep learning models at the edge. The work done by [26] proposes 8-bit quantization to quantize the transformer model for language tasks effectively but with a higher quantization error. Another work by [27] proposed a zero-shot quantization framework for the CNN based model that enables mixed-precision quantization without needing training and validation data. There has also been a focus on the weight rounding technique to quantize the trained model as proposed in the work by Nagel et al. [28]. Recent work done by [29] proposed an effective post-training quantization algorithm that finds optimal low-bit quantization intervals to preserve the relative order of the self-attention results after quantization for the vision transformer model.

Most of the above methods were developed for CNN-based models and were not mainly designed for EEG-related tasks. Furthermore, the approach related to the vision transformer was limited to natural images and hence lacks the quantification of quantization error for transformer-based models trained on the EEG datasets.

3   Methodology

In this section, we summarize the functioning of the Brain Signal Vision Transformer (BSVT); we refer to the work by [5] for full details. Subsequently, we elaborate on the post-training quantization techniques adapted for our experiments.

3.1   Preliminaries on Vision Transformer

Vision Transformer (ViT) [14] takes input the 2D image of height (h), width (w), channels (c), and transforms the image into N non-overlapping $p \times p$ patches, $x^i \in \mathbb{R}^{p \times p}$. The patches are linearly projected by a learned weight matrix $\mathbf{E}$ into a d-dimensional feature vector $\mathbf{z} \in \mathbb{R}^d$. The ViT model adds positional embedding ($\mathbf{E}_{pos}$) to each vector to track its spatial location within the image. It also adds

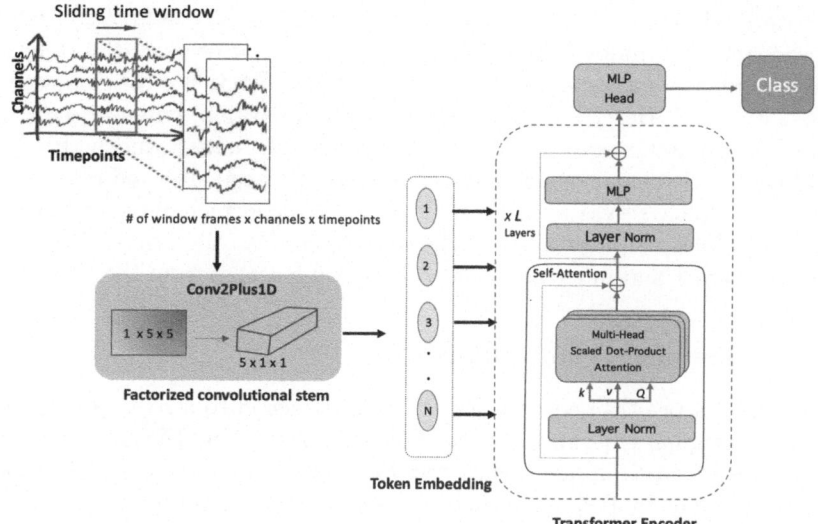

Fig. 2. The BSVT is a transformer-based architecture for classifying EEG signals. It includes the *factorized convolution stem* to extract and embed spatio-temporal features of EEG signals effectively. The factorized convolution stem comprises a 2D convolution followed by a 1D convolution. The figure is based on Khadka et al. [5].

a learned classification embedding (Z_{cls}) to the feature vector, which is then fed into the transformer encoder of L layers [15] with a classification head.

$$\mathbf{z} = [Z_{cls}, \mathbf{E}x_p^1, \mathbf{E}x_p^2..., \mathbf{E}x_p^N] + \mathbf{E}_{pos} \tag{1}$$

where, $\mathbf{E} \in \mathbb{R}^{(p^2c)\times D}$ and $\mathbf{E}_{pos} \in \mathbb{R}^{(N+1)\times D}$.

A transformer encoder block includes Multi-Headed Self-Attention (MSA) [15], layer normalization (LN), and multi-layer perceptron (MLP) blocks as below.

$$\mathbf{y}^l = \mathbf{MSA}(\mathbf{LN}(\mathbf{z}^l)) + \mathbf{z}^l \tag{2}$$

$$\mathbf{z}^{l+1} = \mathbf{MLP}(\mathbf{LN}(\mathbf{y}^l)) + \mathbf{y}^l \tag{3}$$

The term *stem* refers to the part before the transformer encoder block in the literature. The *stem* divides the images into patches, and the process is called *patchifying*.

3.2   Brain Signal Vision Transformer (BSVT)

Our method is based on the Brain Signal Vision Transformer (BSVT) model [5], specifically designed to handle EEG classification tasks. BSVT employs a transformer backbone with a patchifying block that captures the spatial and temporal features of EEG signals and converts them into tokens (see Fig. 2). Given the multivariate EEG signals of a subject, represented as $\mathcal{X} \in \mathbb{R}^{T \times H \times W}$, where T is the sequence of window frames, and H and W denote the spatial dimensions of height and width, respectively, the patchifying block produces embeddings $\tilde{\mathcal{X}} \in \mathbb{R}^{p_t \times p_h \times p_w \times d}$. Here, d represents the token dimension, while p_t, p_h, and p_w are the image patches extracted from the temporal, height, and width dimensions of the EEG tensor, respectively. These embeddings are then reshaped into $\mathbf{X} \in \mathbb{R}^{N \times d}$, which serves as the input to the transformer encoder block. The patchifying block employs a $(2+1)\mathbf{D}$ convolution, as illustrated in Fig. 2, to effectively extract and integrate spatial-temporal features from the EEG signals.

3.3   Post-Training Quantization(PTQ)

PTQ converts a pre-trained network of FP32 precision to a fixed point network without the original training pipeline. This approach offers the advantage of not requiring data for retraining or may require a small calibration set. As reported in the white paper by [10], one crucial step is to find a good quantization range, that significantly affects the quantization error. We take a pre-trained BSVT network and select the quantization parameters for each layer's weight and activation tensors. In setting the quantization range, a clipping threshold of the quantization grid is selected, denoted as p_{min} and p_{max} (see Eq. 4). The other quantization parameters are scaling factors s and zero point z. The Eq. 4 takes in floating-point value x and maps it into an integer value x_{int}. The objective is to minimize the mean squared error (MSE) between the tensors before and after quantization. Unlike weight quantization, determining parameters for activation quantization requires a few batches of calibration data.

$$x_{\text{int}} = \text{clamp}\left(\left[\frac{x}{s}\right] - z, p_{\text{min}}, p_{\text{max}}\right) \tag{4}$$

For approximating the real-valued input x, a de-quantization step is performed :

$$x \approx \hat{x} = s\left(x_{\text{int}} - z\right) \tag{5}$$

Combining Eqs. 4 and 5, the quantization process can be written as :

$$\hat{x} = s\left(\text{clamp}\left(\left[\frac{x}{s}\right] - z, p_{\text{min}}, p_{\text{max}}\right) + z\right) \tag{6}$$

The quantization loss, L_{quant}, is defined as:

$$\arg\min_{s,z} L_{\text{quant}} = D(\hat{x}, x), \tag{7}$$

where $\hat{x}$ is the de-quantized tensor, and D evaluates the distance between $\hat{x}$ and the full-precision tensor x. The distance metric function could include mean squared error (MSE), cosine distance, L_1 distance, and Kullback-Leibler (KL) divergence. The model's weight is quantized by minimizing the quantization error, and the activation tensor is quantized by creating a small calibration

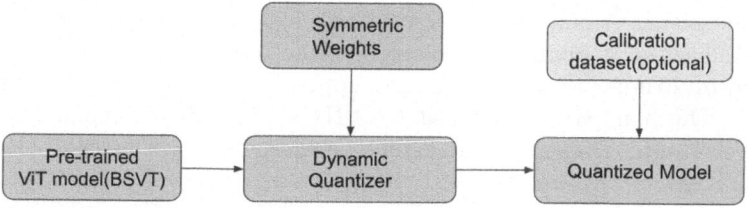

Fig. 3. Illustration of the pipeline for implementing the post-training dynamic quantization. The symmetric weight quantizer is adopted for the weights.

dataset randomly sampled from the training dataset. The scale factor for activation can be determined dynamically based on the observed data per batch during runtime, making the process relatively free of tuning parameters.

4 Experimental Results

In this section, we evaluate post-training quantization on the BSVT model that takes in a 3D tensor as input for classifying EEG signals. We outline the specifics of the implementation, including the dataset utilized for training the BSVT model and the experimental configurations.

4.1 Implementation Details

We focus on implementing the quantization method discussed in Sect. 3.3 for the BSVT model trained on EEG datasets. The pipeline for model quantization is illustrated in Fig. 3. We implemented a quantization version that dynamically determines the scale factor for activations based on the data range observed at runtime. The BSVT model was trained on the EEG dataset, and the `quantize_dynamic` function available in PyTorch was applied. This function takes the model, the list of network layers to be quantized, and the targeted data type for the quantized model.

To obtain the quantized BSVT, the fully connected (FC) layers in BSVT were replaced with quantized layers. For high-precision operations such as Softmax, Layer Normalization, and GELU, the FP32 data type was maintained. The pre-training of the model was conducted on an NVIDIA A100-SXM4 GPU. More detailed information about the model pre-training process can be found in [5]. Quantization and evaluation were performed on a single-threaded CPU as a reference.

Dataset. The model was pre-trained on a publicly available abnormal EEG dataset from the Temple University Hospital (TUH) v2.0.0 [30,31]. This dataset consists of 2,993 recordings from 2,329 subjects. The recordings are labeled as *normal* (1,385 recordings) or *abnormal* (998 recordings). The dataset is divided into a training set of 2,717 recordings and an evaluation set of 276 recordings.

Pre-processing. Among all the recordings from each subject, we selected a common subset of 19 channels. We cropped the recordings of each subject to a maximum of 20 min. Then, signals are clipped at ± 800 µV and downsampled to 100 Hz. The bandpass filtering of 1–40 Hz is applied to attenuate noise. We generate overlapping 5-s segments using the sliding window method (19×500). Thus, the EEG signals are reshaped into a tensor of size $118 \times 19 \times 500$. Finally, we applied the channel-wise normalization to have a zero mean and standard deviation of 1.

Experimental Settings. We design the experimental setup to observe the effects of quantization on the model's performance, latency, and memory footprint. The FP32 model trained on the EEG datasets is quantized to convert it into a fixed-point representation (INT8). To evaluate the effectiveness of the quantization process, we compared the model size before and after quantization. This comparison quantifies the reduction in memory usage achieved through quantization. The computational efficiency of the quantized model is assessed by running inference using the validation dataset. A batch size of 8 is used to handle the memory efficiently. We recorded accuracy and F1 score, before and after the quantization. This assessment shows the trade-offs between reduced model size and the model's performance.

4.2   Results and Analysis

The quantization pipeline was evaluated on the BSVT model trained on EEG datasets, and the experimental results are summarized in Table 1. We compared the model's size before and after quantization. Quantization reduced the non-embedding part of the model's size from 149.00 MB (FP32 model) to 59.72 MB (INT8), achieving an approximate 59% reduction. After quantization, the model's accuracy remained at 0.793, with an F1 score of 0.77, representing an approximate reduction of 2%. Furthermore, we observed that the quantized model (INT8) demonstrated faster inference times. The EEG validation dataset (276 samples) was processed using a single-threaded CPU in 16.98 min, resulting in a 1.47x increase in inference speed.

Quantization results demonstrate a significant improvement in computational efficiency, reducing the model size by 59% and increasing inference speed by 47%. These improvements are particularly impactful in the context of deploying the model on resource-limited edge devices, addressing constraints related to memory and computational power. Furthermore, the trade-off between computational efficiency and performance is minimal, with only a 2% decrease in accuracy and F1 score. This balance between efficiency and performance underscores the effectiveness of the dynamic quantization approach, making the BSVT model suitable for deployment in portable EEG monitoring devices.

Table 1. BSVT model is dynamically quantized into INT8 type. The performance drop after quantization is measured in accuracy and F1 score. The model size is compared before and after quantization. The inference speed is based on a single thread.

Precision type	Accuracy	F1 score	Model Size	Inference time (1 thread)
FP32	0.81	0.79	149 MB	1487.41 s
INT8	0.793	0.77	59.72 MB	1018.70 s

5 Conclusion

We evaluated the impact of quantizing a ViT model (BSVT) designed for EEG datasets by analyzing model size, latency, and performance metrics. The results demonstrate that dynamic quantization is straightforward to implement and results in minimal performance trade-offs. This study highlights the potential of quantization techniques to enhance deep learning tools for EEG analysis, enabling their adoption in resource-constrained environments. In future work, we plan to explore additional model quantization methods to achieve efficient inference while maintaining low latency. Our ultimate goal is to further reduce the memory footprint, enabling the deployment of the model on portable devices for real-world applications.

Acknowledgments. This work has received funding from the European Union's Horizon 2020 research and innovation program under grant agreement No. 964220. We conducted experiments on the Experimental Infrastructure for Exploration of Exascale Computing (eX3) system, financially supported by RCN under contract 270053.

Disclosure of Interests. The authors have no competing interests to declare that they are relevant to the content of this article.

References

1. Ray, P.P., Dash, D., De, D.: Edge computing for Internet of Things: a survey, e-healthcare case study and future direction. J. Netw. Comput. Appl. **140**, 1–22 (2019)
2. Britton, J.W., et al.: Electroencephalography (EEG): an introductory text and atlas of normal and abnormal findings in adults, children, and infants (2016)
3. Georgieva, P., et al.: EEG signal processing for brain–computer interfaces. In: Springer Handbook of Bio-/Neuroinformatics, pp. 797–812 (2014)
4. Wei, Yu., et al.: A survey on the edge computing for the Internet of Things. IEEE Access **6**, 6900–6919 (2017)
5. Khadka, R., et al.: Inducing inductive bias in vision transformer for EEG classification. In: ICASSP 2024-2024 IEEE International Conference on Acoustics, Speech and Signal Processing (ICASSP), pp. 2096–2100. IEEE (2024)
6. Arjun, A., Rajpoot, A.S., Panicker, M.R.: Introducing attention mechanism for eeg signals: emotion recognition with vision transformers. In: 2021 43rd Annual International Conference of the IEEE Engineering in Medicine & Biology Society (EMBC), pp. 5723–5726. IEEE (2021)

7. Al-Quraishi, M.S., et al.: Decoding the user's movements preparation from EEG signals using vision transformer architecture. IEEE Access **10**, 109446–109459 (2022)
8. Nag, S., et al.: ViTA: a vision transformer inference accelerator for edge applications. In: 2023 IEEE International Symposium on Circuits and Systems (ISCAS), pp. 1–5. IEEE (2023)
9. Banner, R., Nahshan, Y., Soudry, D.: Post training 4-bit quantization of convolutional networks for rapid-deployment. Adv. Neural Inf. Process. Syst. **32**, 1–9 (2019)
10. Nagel, M., et al.: A white paper on neural network quantization. arXiv preprint arXiv:2106.08295 (2021)
11. Smith, F.W., Smith, M.L.: Decoding the dynamic representation of facial expressions of emotion in explicit and incidental tasks. Neuroimage **195**, 261–271 (2019)
12. Acharya, U.R., et al.: Automated EEG-based screening of depression using deep convolutional neural network. Comput. Methods Prog. Biomed. **161**, 103–113 (2018)
13. Aggarwal, S., Chugh, N.: Review of machine learning techniques for eeg based brain computer interface. Arch. Comput. Methods Eng. **29**(5), 3001–3020 (2022)
14. Dosovitskiy, A., et al.: An image is worth 16x16 words: transformers for image recognition at scale. arXiv preprint arXiv:2010.11929 (2020)
15. Vaswani, A., et al.: Attention is all you need. Adv. Neural Inf. Process. Syst. **30** (2017)
16. Touvron, H., et al.: Training data-efficient image transformers & distillation through attention. In: International Conference on Machine Learning, pp. 10347–10357. PMLR (2021)
17. Strudel, R., et al.: Segmenter: transformer for semantic segmentation. In: Proceedings of the IEEE/CVF International Conference on Computer Vision, pp. 7262–7272 (2021)
18. Zhang, Z., et al.: ViT-YOLO: transformer-based YOLO for object detection. In: Proceedings of the IEEE/CVF International Conference on Computer Vision, pp. 2799–2808 (2021)
19. Lee, Y.E., Lee, S.H.: EEG-transformer: self-attention from transformer architecture for decoding EEG of imagined speech. In: 2022 10th International Winter Conference on Brain-Computer Interface (BCI), pp. 1–4. IEEE (2022)
20. Zhang, L.L., et al.: Nn-meter: towards accurate latency prediction of deep-learning model inference on diverse edge devices. In: Proceedings of the 19th Annual International Conference on Mobile Systems, Applications, and Services, pp. 81–93 (2021)
21. Hashemi, S., et al.: Understanding the impact of precision quantization on the accuracy and energy of neural networks. In: Design, Automation & Test in Europe Conference & Exhibition (DATE), pp. 1474–1479. IEEE (2017)
22. Fang, W.-C., Wang, K.-Y., Fahier, N., Ho, Y.-L., Huang, Y.-D.: Development and validation of an eeg-based real-time emotion recognition system using edge ai computing platform with convolutional neural network system-on-chip design. IEEE J. Emerg. Sel. Topics Circ. Syst. **9**(4), 645–657 (2019)
23. Gómez-Quintana, S., et al.: An EEG analysis framework through AI and sonification on low power IoT edge devices. In: 2021 43rd Annual International Conference of the IEEE Engineering in Medicine & Biology Society (EMBC), pp. 277–280. IEEE (2021)
24. Saini, M., Satija, U.: On-device implementation for deep-learning-based cognitive activity prediction. IEEE Sens. Lett. **6**(4), 1–4 (2022)

25. Tazrin, T., et al.: LiHEA: migrating EEG analytics to ultra-edge IoT devices with logic-in-headbands. IEEE Access **9**, 138834–138848 (2021)
26. Shen, S., et al.: Q-bert: hessian based ultra low precision quantization of bert. In Proceedings of the AAAI Conference on Artificial Intelligence, vol. 34, pp. 8815–8821 (2020)
27. Cai, Y., et al.: Zeroq: a novel zero shot quantization framework. In: Proceedings of the IEEE/CVF Conference on Computer Vision and Pattern Recognition, pp. 13169–13178 (2020)
28. Nagel, M., et al.: Up or down? adaptive rounding for post-training quantization. In: International Conference on Machine Learning, pp. 7197–7206. PMLR (2020)
29. Liu, Z., Wang, Y., Han, K., Zhang, W., Ma, S., Gao, W.: Post-training quantization for vision transformer. Adv. Neural. Inf. Process. Syst. **34**, 28092–28103 (2021)
30. López, S., Obeid, I., Picone, J.: Automated interpretation of abnormal adult electroencephalograms. PhD thesis (2017)
31. Obeid, I., Picone, J.: The temple university hospital eeg data corpus. Front. Neurosci. **10**, 196 (2016)

6G-Enabled Network Security in Drone Technology: Revolutionizing Communications and Enhancing Security Applications

Sabiul Islam[1]([✉]), Andrew Karam[2], Adda Boualem[3], Khaja Valli Pathan[4],
Ahmed F. Sayed[5], and Atta-ur Rahman[6]

[1] NCS Department, SUNY POLY, Utica, NY 13502, USA
234sabiul@gmail.com
[2] The Air Force Research Laboratory (AFRL, RIGB), Rome, NY, USA
andrew.karam@us.af.mil
[3] Department of Science Computer, Ibn Khaldoun University, Tiaret, Algeria
adda.boualem@univ-tiaret.dz
[4] CIS Department, SUNY, Albany, USA
[5] Transmission Department, Telecom Egypt, Fayoum, Egypt
[6] Department of Computer Science (CS), College of Computer Science and Information
Technology(CCSIT), Imam Abdulrahman Bin Faisal University (IAU), P.O. Box 1982,
Dammam 31441, Saudi Arabia

Abstract. The combination of 6G technology and drones represents a major achievement in enhancing the use of unmanned aerial vehicles (UAVs). This article investigates how 6G can revolutionize communication and security systems for drones in various sectors such as disaster management, surveillance, logistics, and infrastructure monitoring. 6Gpowered drones improve processing data in real-time, secure communication, and autonomous decision-making abilities by utilizing advanced technologies like terahertz communication, blockchain, AI-driven threat detection, and federated learning. The research also examines the urgent cybersecurity issues linked to drone networks, like denial-of-service attacks, unauthorized entry, and data breaches, by suggesting complex structures and cutting-edge machine-learning methods for dealing with threats. This research showcases the extensive capabilities of drone systems enabled by 6G, while also pointing out important constraints in scalability, energy efficiency, and standardization, offering a guide for further investigation. The results highlight the crucial impact of 6G on transforming UAV operations, and creating secure, robust, and flexible networks for civilian and military uses.

Keywords: 6G technology · drone networks · cybersecurity · blockchain · terahertz communication · unmanned aerial vehicles (UAVs) · threat detection · Internet of Things (IoT) · secure communication · autonomous decision-making

© The Author(s) 2025
A. Azab and T. Malkiewicz (Eds.): NeIC 2024, CCIS 2398, pp. 28–44, 2025.
https://doi.org/10.1007/978-3-031-86240-3_3

1 Introduction

The rapid advancements in 6G technology and its integration with Internet of Things (IoT) ecosystems have revolutionized unmanned aerial vehicle (UAV) applications across various domains. From disaster management and surveillance to logistics and infrastructure monitoring, drones powered by 6GIoT connectivity offer unprecedented capabilities in real-time data processing, communication, and autonomous decisionmaking. However, this growth comes with significant cybersecurity challenges, as drones are increasingly susceptible to attacks such as denial-of-service (DoS), remote-to-local (R2L), unauthorized access, and data breaches. These vulnerabilities, if unaddressed, can compromise critical infrastructure and endanger sensitive operations [2]. New frameworks are using advanced technologies like blockchain, AI, and ML to improve drone network security and privacy to reduce these dangers. In highly networked environments, multi-layered architectures that include sophisticated threat detection and response systems are intended to protect drone communications and data integrity. This study investigates how 6G drone technology can be used to address these security issues, offering creative solutions that improve cyber-threat resistance in both the military and civilian sectors.

2 Research Background

The fast development of sixth-generation (6G) communication technology presents new chances for enhancing unmanned aerial vehicle (UAV) systems. Drones are being used more frequently in various fields like precision agriculture, smart city infrastructure, disaster response, and military operations, as key elements of current networks. Their capability to serve as middlemen connecting land-based and sky-based networks allows for smooth communication, backing up quick data transmission, very little delay, and widespread coverage. Nevertheless, the distinctive features of UAVs, including their high mobility, restricted energy sources, and reliance on line-of-sight communication, create notable obstacles in attaining dependable and secure network integration.

The increasing use of UAVs requires new ways to combat cybersecurity risks like denial-of-service attacks, data breaches, and jamming that endanger critical infrastructure and sensitive operations. Utilizing 6G technologies like terahertz communication, massive MIMO, and free-space optics, along with advancements in blockchain and machine learning, presents exciting opportunities to improve drone network security and performance. This part explores the latest progress and obstacles in merging 6G with UAVs, setting the groundwork for creating robust and effective drone communication systems.

Mishra et al. (2021) [1] In this study, the revolutionary potential of incorporating Unmanned Aerial Vehicles (UAVs) into sixth-generation (6G) communication networks is examined. It highlights how UAVs play a crucial role as bridges between terrestrial and space networks, creating a smooth foundation for space-air-ground communication to meet 6G demands including high data speeds, ultra-low latency, and widespread coverage. The report identifies a number of enabling technologies that enable effective communication between ground, aerial, and space sectors, such as massive MIMO, terahertz (THz), millimeter-wave (mmWave), and free-space optics (FSO). Challenges like guaranteeing secure communication and negotiating regulatory frameworks are also explored, along with important design factors like energy consumption, spectrum management, and mobility support. By concentrating on a variety of applications, including precision agriculture, smart city infrastructure, and crisis management, the report offers insightful information about how UAV-based systems might improve 6G networks. Despite providing a clear overview of the technological possibilities, the study highlights the need for more research and cooperation in this developing field by recognizing significant obstacles in standardization, implementation, and real-world deployment. The study assesses design factors such as cross-layer optimization, energy-efficient routing, and smooth handovers to guarantee dependable and economical network performance. One of the paper's main strengths is its identification of issues, such as spectrum sharing, legal restrictions, and UAVs' susceptibility to security risks including jamming and eavesdropping. By highlighting the necessity of specialized UAV-specific protocols to synchronize aerial and terrestrial systems, the authors also address the standardization initiatives by organizations such as 3GPP, IEEE, and ITU. Ultimately, the paper addresses the fundamental obstacles that need to be removed in order to achieve this objective while highlighting the revolutionary influence of UAVs on 6G networks.

Tandra et al. (2024) [2] A seven-layered architecture combined with 6G IoT technology and sophisticated machine learning models is the unique method for tackling privacy and cybersecurity issues in drone networks that is presented in this research. The Regression Net ensemble model, which combines Logistic Regression and Multilayer Perceptron (MLP), is part of the suggested framework. It achieves an excellent 99.89% intrusion detection accuracy. Resilience is demonstrated by validation on datasets including STIN and KDD CUP 99, and it successfully classifies threats like DoS, R2L, U2R, and Prob assaults. According to the report, modern sensors are used for real-time anomaly detection and effective drone operation, while blockchain and the Internet of Things are used for safe communication and data storage. Its contributions to the guarantee of safe drone operations in both military and civilian settings are substantial. The paper's shortcomings, however, include its reliance on pre-existing datasets that might not accurately reflect new attack scenarios and its omission of any remarks regarding scalability and legacy system integration. These holes could be filled by future studies that curate fresh datasets, examine scalability in massive networks, and look at regulatory frameworks for safe deployments. This work contributes significantly to the field of safe IoT-enabled drone systems by offering perspectives for upcoming developments in drone cybersecurity.

Hakeem et al. (2022) [3] The new 6G wireless network, which is expected to transform mobile communications by 2030, is examined in this study. With innovations like blockchain, AI, machine learning, visible light, terahertz, and molecular communication, 6G promises improved connection, low latency, and energy efficiency. Nevertheless, these developments call for a review of conventional security protocols. To achieve 6G's strict security, privacy, and trust requirements, the study emphasizes the necessity of innovative authentication, encryption, and harmful activity detection techniques. With an emphasis on technologies like edge computing and quantum communication, a thorough examination of the security evolution from 1G to 5G, as well as the special security architecture and difficulties of 6G, is provided. These technologies each have serious security flaws in addition to their revolutionary promise. Critical applications, like increased low-latency communication and the Internet of Everything, are further examined in the article along with the security issues they raise. Decentralized security systems, post-quantum cryptography, and zero-trust architecture are some of the solutions put out to deal with problems brought on by resource limitations, distributed networks, and high mobility. Integration of AI/ML also creates new hazards, but it also provides avenues for dynamic threat identification and response. The study highlights the importance of taking preventative action to secure 6G and offers a comprehensive strategy to address both known and unknown threats. It provides a basis for future studies on the security architecture of 6G networks and the application of innovative technologies in real-world settings.

Al-Turjman et al. (2022) [4] To solve issues in environmental catastrophe management, this study investigates the integration of fog computing with unmanned aerial vehicles (UAVs) in the Internet of Everything (IoE)-based systems. UAVs are emphasized for their versatility, mobility, and capacity to collect data in real-time in remote and disaster-prone locations. In crucial situations, fog computing is stressed for its ability to process data closest to the source, reduce latency, and facilitate quicker decision-making. When combined, these technologies offer a strong framework for improving efforts at disaster response, recovery, and prediction, especially in situations where conventional infrastructure can malfunction. To maximize the effectiveness of communication, scalability, and resource allocation during disasters, the study suggests a paradigm that integrates UAVs with fog computing. Comprehensive case studies that illustrate the system's real-world implementation are included, highlighting advancements in resource coordination, data collection, and quick reaction. Through tackling significant technical obstacles like data processing, network dependability, and system adaptability, the study highlights the revolutionary possibilities of incorporating UAVs and cloud computing into the Internet of Everything systems to create robust infrastructures for disaster recovery and prediction. Smart catastrophe management techniques are advanced by this work, opening the door for more study and possible applications in this field.

Deebak et al. (2023) [5] Utilizing cutting-edge communication technologies such as Beyond 5G (B5G) and 6G, this study discusses the incorporation of UAVs, or unmanned aerial vehicles, into military intelligence systems. The authors suggest a strong lightweight secured multi-factor identification (RL-SMFA) system that uses AI to improve secure communication and authentication procedures. The RL-SMFA's lower power consumption, mutual authentication, and data confidentiality are essential for drone networks with limited resources. Through simulations and formal verification, the system surpasses current approaches in terms of performance, delay from end to end, and the packet delivery ratio and shows resilience against possible threats like denial of service attacks and data breaches. To verify drone authenticity and create a safe connection with ground control systems, the suggested method makes use of elliptic-curve digital encryption and an assisted by AI secure analytics phase. Additionally, it tackles important security and performance issues like untraceability, session key agreement, and user anonymity. The paper emphasizes how the RL-SMFA may be used in both military and medical settings, providing a scalable and energy-efficient framework for monitoring and managing data in real-time. The implementation of safe UAV-based systems in rapidly evolving and dangerous environments is made possible. Angjo et al. (2021) [6] The quick advancement of drone technology, particularly their incorporation into upcoming 6G mobile networks, has brought about hopeful solutions for improving communication reliability and coverage in various settings. Drones, used as mobile base stations or connected user devices, are especially beneficial in disaster situations and crowded areas where traditional networks have restrictions. Yet, the distinct ability to move in a three-dimensional (3D) area, dependence on line-of-sight (LoS) communication, and fast movement make handover (HO) management much more challenging. The need for innovative strategies arises due to challenges like frequent handovers, interference from Line of Sight paths, and high packet error rates to ensure continuous connectivity and uphold Quality of Service (QoS). Previous studies have investigated algorithmic, machine learning, and stochastic geometry methods to improve HO procedures and address problems such as false HO triggers, coverage gaps, and signal interference.

This research provides a thorough examination of handover control for drones in 6G mixed networks, focusing on major difficulties and suggested remedies. Sophisticated methods such as deep learning and mobility prediction algorithms can tackle the ever- changing drone movement and connectivity requirements. Moreover, the document highlights the significance of integrating innovative methods like coordinated multi- point (CoMP) transmission and dynamic beamforming to improve drone communication. Even though there have been advancements, the research report points out shortcomings in current studies, like adjusting protocols for drones' unique movements and enhancing HO mechanisms for ultra-dense and high-speed situations. These observations will serve as the foundation for upcoming advancements in incorporating drones into advanced wireless networks.

Shvetsov et al. (2023) [7] This research paper examines the combination of Federated Learning (FL) and Intelligent Reflecting Surfaces (IRS) in drones, offering a groundbreaking method for enhancing 6G wireless networks. The suggested approach utilizes drones equipped with IRS to improve wireless communication dynamically through signal reflection optimization and optimized drone paths. Through the integration of FL, a decentralized approach to machine learning, with IRS, the system can achieve flexible and effective communication solutions. Employing drones equipped with IRS allows for increased coverage, faster data rates, and energy-efficient operations, making them well-suited for various uses such as disaster relief, IoT, and connecting remote areas. Some advantages consist of lower energy usage, improved spectrum efficiency, and the capability to overcome obstructions in direct communication. Furthermore, this study observes the technical obstacles and potential benefits of this merger. Emphasizing the importance of optimizing drone flight paths and IRS setups to handle intricate and ever-changing network surroundings. The writers suggest new ideas to tackle complex optimization issues, enhance Quality of Service (QoS), and bolster network security from eavesdropping risks. By utilizing simulations and analytical conversations, the study highlights how FL and IRS could transform aerial communication by providing scalable, secure, and efficient networking features. This study provides a base for future exploration of 6G technologies, highlighting the significance of integrating advanced technologies to address evolving communication needs. Amponis et al. (2022) [8] The authors investigate the incorporation of drones as airborne base stations in Beyond 5G (B5G) and 6G networks, highlighting their ability to transform network connectivity by providing better flexibility and mobility. It examines cutting-edge technologies like millimeter-wave communications, MIMO systems, and AI integration that support the needs of future IoT applications. The writers emphasize the benefits of drones, such as quick deployment, flexibility, and their capacity to conquer geographical and infrastructural barriers. Specific uses involve expanding network range, aiding mobile ad hoc networks, and using beamforming to handle interference in crowded city settings. The examination also looks into the difficulties in this situation, like limited energy, best positioning, movement, safety, and quality of service, suggesting methods to optimize and new structures to tackle them. This research analyzes examples and upcoming trends, such as the utilization of tethered drones for extended missions and the implementation of AI-based algorithms for coordinating swarms and optimizing deployment. It presents a plan to combine drone- related systems with essential network features to guarantee smooth, safe, and easily expandable operations. The results of the study are part of the 5G-INDUCE project, showing how sophisticated orchestration methods can improve network dependability and flexibility in important situations such as disaster response and smart agriculture.

These findings offer a detailed plan for implementing drone-supported cellular networks in the changing 6G environment, overcoming existing challenges and paving the way for a stronger and more interconnected network system.

Abu et al. (2023) [9] This research investigates combining software-defined networking (SDN) with unmanned aerial vehicles (UAVs) to tackle the obstacles of future 6G networks. This creative blend is designed to address the drawbacks present in conventional UAV-operated cellular networks, including centralized monitoring, resource allocation, and proactive management. SDUAV networks, enabled by.

SDN, offers improved flexibility, scalability, reliability, and connectivity, crucial for 6G services such as URLLC, uMUB, and uHDD. The article highlights how SDN's ability to be programmed and centrally controlled is advantageous in overseeing UAV networks, allowing for advanced capabilities like intelligence, automation, and adaptiveness. Main difficulties such as security, interoperability, and spectrum management are addressed, along with emerging technologies like AI, blockchain, and edge computing that can improve network intelligence and effectiveness.

The authors offer a detailed examination of SDUAV structure and communication methods, demonstrating its capacity to enhance 6G services with limited reliance on infrastructure. They pinpoint important deficiencies in current research, such as improving UAV positioning, strengthening security measures, and creating efficient deployment algorithms. The study showcases the ability of SDUAV networks to meet the diverse and changing requirements of 6G networks by facilitating smooth movement, efficient resource allocation, and fast communication. Furthermore, the article proposes potential areas for future research, such as investigating the incorporation of cutting-edge technologies such as quantum communication, proactive caching, and intelligent reflecting surfaces to enhance SDUAV network efficiency. The paper offers a useful guide for improving UAV networking technologies to address the challenges of 6G wireless communication systems.

Zahid et al. (2024) [10] This research paper introduces a Composite Ensemble Learning (CEL) model created for categorizing drones using their Radio Frequency (RF) fingerprints in the scope of upcoming Sixth-Generation (6G) networks. The framework focuses on enhancing RF signal classification to tackle security threats from unauthorized drones. Major advancements involve combining wavelet based signal denoising with automatic and manual feature extraction methods, improving the strength and variety of feature representation. The suggested technique showed better classification results in different Signal-to-Noise Ratios (SNRs) compared to current deep learning standards, particularly in noisy situations.

The approach utilizes cleaned RF signals to extract features through CNNs, LSTM networks, and manually crafted derivatives, which are combined into a composite feature vector for classification. Results from experiments conducted on a publicly available dataset confirmed that the model is efficient, demonstrating high accuracy even in situations with low signal-to-noise ratio. The CEL method showed stability and resilience in noisy environments, indicating its ability to secure drone operations in 6G networks with critical dynamic spectrum sharing and mobility management needs. This study emphasizes the significance of combining traditional signal processing with cutting-edge deep learning for analyzing RF signals. The CEL framework tackles both practical issues in identifying passive drones and provides scalability for 6G-era applications. It offers major improvements in wireless security, which can optimize resource allocation, improve airspace safety, and oversee various drone applications in civilian and military areas. Mahmoud et al. (2021) [11] This research extensively explores the anticipated sixth-generation (6G) cellular networks. The authors discuss the technological constraints and difficulties present 5G networks face, emphasizing the necessity for advancements to accommodate upcoming applications such as extended reality (XR), wireless brain computer interactions (BCI), and autonomous systems. They describe the possibilities of 6G to provide unparalleled data speeds, reliable communication with low latency, and connection with IoT networks in urban areas and beyond. The research classifies developments in various areas, such as terahertz spectrum usage, AI-powered network operation, intelligent coatings for communication, and blockchain for improved security. The study explains how 6G will combine different technologies, such as edge computing, quantum communication, and intelligent surfaces, to offer custom connectivity solutions and enhance energy efficiency. Furthermore, ongoing global research efforts and critical open challenges in 6G development are highlighted in the paper, including the need for ultra-low latency, higher spectral efficiency, and addressing trust, security, and privacy issues. It presents new service categories such as MBRLLC and mURLLC designed for upcoming applications. Furthermore, the research suggests a plan highlighting cross-disciplinary methods, such as utilizing AI for enhancing network efficiency and implementing quantum encryption for ensuring security. This survey creates a foundational framework for researchers and industry stakeholders by summarizing current efforts and future requirements, paving the way for innovative 6G solutions in line with the vision of a connected and smart world.

Stavroulaki et al. (2021) [12] The focus of the DEDICAT 6G project is to tackle the obstacles of future communication networks by creating new ideas for flexible connectivity, decentralized intelligence, and improved interactions between humans and machines. The article details a holistic plan for 6G networks, highlighting the incorporation of artificial intelligence (AI), edge computing, and blockchain technologies to establish a flexible, high-speed, and protected communication infrastructure. The project's main goal is to allow for flexible spreading of computational intelligence in order to enhance resource allocation, increase energy efficiency, and reduce latency. Moreover, the study explores the expansion of coverage in real-time by utilizing mobile access points like drones, robots, and connected vehicles to maintain connectivity in remote locations or during times of crisis. The project relies on AI-powered threat detection and blockchain-based trust management to ensure security, privacy, and trust.

The real-world uses of DEDICAT 6G are demonstrated through four scenarios: Smart Warehousing, Enhanced Experience, Public Safety, and Smart Highways. These examples showcase the usefulness of distributed intelligence, dynamic coverage, and advanced interfaces such as augmented reality (AR) devices. One example is the Smart Warehousing use case, which combines automated guided vehicles (AGVs) and IoT systems to improve operations. In contrast, the Smart Highway scenario prioritizes ultra-low-latency communications for autonomous vehicles and safety upgrades. The project focuses on testing in real-world settings and receiving feedback from stakeholders to confirm the technology's readiness for market launch.

Chang et al. (2023) [13] This research examines the growing significance of UAV-based flying base stations (UxNB) in 5G and 6G networks. These mobile platforms play a crucial role in expanding connectivity, particularly in rural areas or emergencies, utilizing their adaptability for ideal placement and enhanced line-of-sight communication in mmWave channels. Nevertheless, their ability to move around brings about distinct weaknesses not found in conventional stationary land systems. The analysis points out important security issues like managing mobility, dealing with battery constraints, and coordinating distributed tasks, all of which can be targeted by denial-of-service (DoS) attacks, jamming, and GPS spoofing. It stresses the importance of using integrated systems approaches to tackle these issues and ensure the operational integrity and availability of UAV systems, which are essential for dependable mobile connectivity. Moreover, the paper points out deficiencies in existing research, especially in protecting the digital, physical, and dispersed components of UAV base stations. The paper suggests implementing"security by design" in standardization processes by utilizing advancements in technologies such as cryptographic protocols, software- defined networking, and battery management. It also explores the potential advantages that UAVs provide in enhancing network reliability and physical-layer security through their mobility. Plans involve creating functional models, utilizing blockchain technology for managing keys across multiple locations, and improving the establishment of secure communication protocols for drones in the 6G platform. The authors hope to promote more research to confirm these aerial base stations as safe and essential parts of upcoming mobile networks.

Song et al. (2023) [14] This research analyzes aerial platform technology-based wireless backhaul networks as a possible remedy for upcoming 6G communication systems. Aerial platforms, which are divided into Low Altitude Platforms (LAPs) and High Altitude Platforms (HAPs), are offered as adaptable and affordable substitutes for the current backhaul infrastructures due to the growing need for high data rates and their constraints. While HAPs, such as airships and balloons, cover larger regions and have longer operational endurance, LAPs, such as UAVs and drones, offer agility for small-scale deployments. The paper emphasizes how Free Space Optics (FSO) communication might help traditional RF-based systems overcome obstacles. It goes into detail about design concerns such deployment methods, channel modeling for various atmospheric circumstances, energy efficiency, and security challenges. The study also emphasizes the use of technologies such as Software-Defined Networking (SDN) and Multi-Access Edge Computing (MEC) to improve the flexibility and computational functions of aerial networks. It deals with addressing obstacles in optimizing deployment, predicting traffic, and securing communications from cyber threats by suggesting creative algorithms and frameworks to enhance system reliability and efficiency. The study highlights the possibilities of aerial platforms as a foundation for future networks by exploring theoretical perspectives, real-world uses, and example situations, providing recommendations for overcoming key obstacles in network scalability and performance enhancement.

Saad et al. (2019) [15] This research paper details a holistic plan for 6G wireless networks, focusing on advancing past the constraints of 5G to meet the demands of new applications such as multisensory extended reality (XR), connected robotics, wireless brain-computer interactions, and blockchain technologies. The writers suggest that 6G will prioritize combining various technologies to enable fast data speeds, minimal delays, and reliable connections, while also supporting services centered around humans and distributed environments. They suggest introducing new types of services like MBR-LLC, mURLLC, and HCS, which redefine old wireless service standards. Important developments in 6G include smart surfaces becoming prevalent, advancements in spectral and energy efficiency, integration of communication with control and sensing, and a transition from smartphones to implantable devices.

This paper highlights important technologies needed for 6G to become a reality, such as progress in terahertz frequencies, large intelligent surfaces (LIS), edge AI, and the merging of terrestrial, airborne, and satellite networks. It emphasizes the importance of new 3D networking protocols, system designs driven by AI, and the combined optimization of communication, computation, and control features. Furthermore, it outlines a research plan aimed at addressing key issues like balancing rates, reliability, and latency, developing new metrics for quality of physical experience (QoPE), and integrating radio frequency with non-radio frequency connections. In their conclusion, the authors suggest a significant change in wireless system design, influenced by innovative applications and technical advancements.

Bandi et al. (2022) [16] The security and privacy issues of 6G wireless networks are examined in this research, along with potential solutions. The authors categorize the privacy and security issues unique to AI-powered apps under 6G technology. Among the main risks are encryption problems, machine learning API flaws, and denial of service (DoS) assaults. To reduce these dangers, the study also looks at defense strategies such as biometric identification, explainable AI, homomorphic encryption, and trustworthy AI. Terahertz (THz) frequencies, blockchain, and quantum communication are some of the new technologies that the report highlights as having a significant impact. Big data analytics, human-centric apps, autonomous systems, and other categories are used by the taxonomy to group security and privacy issues. The study focuses on how applications such as extended reality, autonomous vehicles, and the Global Internet of

Everything (IoE) may be affected by weaknesses in AI/ML models, connection problems, and physical layer attacks. Utilizing physical layer authentication, superintelligent AI, and cuttingedge encryption technology are some of the defensive tactics suggested. This work adds to the fundamental knowledge needed to handle the complex threats anticipated in nextgeneration mobile networks while guaranteeing reliable and secure 6G application implementations.

Puspitasari et al. (2023) [17] This study highlights the importance of machine learning (ML) while reviewing the problems and possible fixes for 6G communication networks. In light of the unprecedented demands of 6G, including massive connectivity, ultra- reliable low-latency communication, and extreme energy and spectral efficiency, the paper highlights important emerging technologies such as non-orthogonal multiple access (NOMA), Intelligent Reflecting Surfaces (IRS), and unmanned aerial vehicles (UAVs). It draws attention to the drawbacks of conventional mathematical approaches to technology optimization and makes the case for the application of machine learning (ML) methods, such as the use of deep learning (DL) and reinforcement learning (RL), to get over these obstacles. The authors go into detail on how these algorithms can improve system performance, solve important performance metrics for 6G networks, and address computational complexity. A thorough analysis of several machine learning models applied to new 6G technologies is given in the study. These include IRS for enhancing signal strength and spectral effectiveness, UAVs for wireless communication deployment flexibility, and NOMA for effective spectrum utilization. It looks at how ML may help with phase shift configurations, trajectory planning, and power allocation optimization, among other things. The study concludes that combining ML with these advancements is crucial to realizing the intended objectives of 6G networks.

Furthermore, the authors delineate unresolved issues such as data privacy, computing overhead, and system coordination and propose avenues for further investigation into the utilization of machine learning for future communication systems.

Uwaoma et al. (2023) [18] This research study examines the security issues and solutions related to the implementation of 6G technologies within healthcare systems, highlighting the revolutionary possibilities of 6G features including blockchain-enabled data security, distributed artificial intelligence, and real-time intelligent edge computing. These developments bring serious security flaws carried over from previous network generations, even while they hold promise for innovations in wearable medical technology, medical drones, and remote surgery. The paper suggests a Zero Trust framework combined with 6G-specific elements including decentralized blockchain technology, quantum cryptography, and AI-driven security protocols to allay these worries. To guarantee the safe implementation of 6G networks in healthcare and improve the quality of services and data protection, the study emphasizes the necessity of proactive security measures.

Kishk et al. (2020) [19] The use of tethered drones (tUAVs) for the deployment of airborne base stations (BS) for 6G cellular networks is covered in the article. The authors suggest this configuration to get around the drawbacks of untethered drones (uUAVs), namely their short flight durations brought on by low onboard battery and erratic wireless backhaul connections. UAVs can operate continuously for long periods thanks to a tether that connects them to a ground station and provides energy and dependable data connectivity. Longer flight duration, more payload capacity for improved communication and interference control, and increased reliability are some of the benefits of tUAVs highlighted in the paper. Use cases covered include capacity building in crowded urban areas and coverage in rural or disasteraffected areas. The study also looks into the trade-offs between durability and mobility, highlighting the fact that although tUAVs have somewhat less mobility because of tether limitations, they nonetheless offer reliable, highquality communication on par with terrestrial BSs. The better coverage chances of tUAVs over uUAVs under a variety of operational conditions are shown via Monte Carlo simulations. The authors list difficulties, including tether length restrictions, the best locations for ground stations, and regulatory issues, and they suggest research avenues for the future to overcome these restraints. The study presents tUAVs as a possible solution to improve next-generation wireless networks' scalability and dependability.

Akinbi et al. (2023) [20] The paper examines how sixthgeneration (6G) wireless communication technology can revolutionize IoT networks by providing high speeds, reduced latency, and combined terrestrial, space, and underwater connections. It focuses on important 6G uses in extended reality, autonomous systems, and smart cities, while also discussing the new digital forensic issues that arise from these developments. These encompass challenges in gathering, protecting, and examining evidence in complex, diverse settings with large amounts of data. The research also examines forensic preparedness by suggesting possible future research paths, such as creating specialized methodologies and tools for 6G-enabled IoT settings, stressing the importance of edge intelligence, XR/VR forensics, and optical wireless networks.

Ning et al. (2023) [21] This study offers a thorough examination of UAV communications with the assistance of Intelligent Reflective Surfaces in the framework of 6G networks. It examines how IRS and UAV technologies can tackle issues like limited coverage, high energy usage, and changing communication environments in 6G systems. The article presents the main problems, technologies, and use cases of UAV communication assisted by the IRS, such as energy-limited and secure communication options. It elaborates on creative applications, including improving Space-Air-Ground Integrated Networks (SAGIN), facilitating Vehicle-to-Everything (V2X) communications, and backing large-scale IoT networks. Ultimately, it focuses on unresolved issues and upcoming research areas, providing valuable perspectives for enhancing integrated IRS-UAV systems in future wireless networks.

Wang et al. (2024) [22] The integration of unmanned autonomous intelligence systems (UAIS) with 6G nonterrestrial networks (NTN) is examined in this paper, with a focus on how these systems can facilitate advanced communication in situations when conventional infrastructure is inadequate. It examines how NTN technologies, such as satellites, unmanned aerial vehicles, and high-altitude platforms, contribute to dependable and ultra-low latency communication, which is essential for applications like robotics, driverless cars, and disaster relief. Presenting case examples to highlight field experiments and possible applications, the paper draws attention to issues like communication dependability, energy constraints, and edge computing requirements. It emphasizes the need for distributed computing architectures to offload large amounts of data and the revolutionary potential of 6G NTN to enable AI-driven UAIS in challenging and remote settings. Masaracchia et al. (2021) [23] This research investigates how unmanned aerial vehicles (UAVs) can be incorporated into ultra- reliable low-latency communication (URLLC) networks to support 6G. It starts by emphasizing the important characteristics and difficulties of URLLC, including strict demands for latency, reliability, and energy efficiency, and talks about how UAVs can improve network coverage and quality. The research study offers an in-depth categorization and examination of the latest advancements in UAV-supported URLLC systems, discussing important topics like UAV deployment, power management, block length improvement, and cross-layer strategies. It highlights the capability of UAVs to enhance performance in areas such as industrial automation, telemedicine, and VR/AR while addressing remaining challenges like widespread deployment, immediate execution, and security concerns. The paper ends by providing insights on potential future paths for UAV-enabled URLLC systems.

Wang et al. (2020) [24] This research extensively investigates the new needs, metrics for performance, and technologies that support 6G communication networks. It outlines progress in different areas such as mmWave, THz, and optical wireless, and also includes scenarios like satellite, UAV, maritime, and underwater communication. Ultra- massive MIMO, vehicle-tovehicle (V2V), and industrial IoT channels are some of the specific areas of focus. The writers point out difficulties in channel modeling, stressing the importance of a unified 6G framework that includes AI for dynamic channel measurement and modeling. They support studying Intelligent Reflecting Surfaces (IRS) and hybrid models to tackle challenges in various high-frequency mobility scenarios.

Deebak et al. (2020) [25] This research investigates the combining of drones with IoT systems in the context of upcoming 6G networks. It emphasizes the capabilities of 6G in increasing data transmission speeds, boosting system capacity, and enhancing service quality for a wide range of applications including smart agriculture, automated home systems, and smart metering. The research highlights the important role of artificial intelligence in enhancing system intelligence and UAV capabilities. Moreover, it delves into merging UAV and satellite systems as a possible resolution for overcoming technological restrictions, integration hurdles, and network intricacies in IoT deployments, especially in smart cities. Further studies that focus on the cybersecurity aspects are given in [26–72].

3   Conclusion and Future Work

Incorporating 6G technology into drone systems is a significant advancement in improving communication, security, and efficiency in various uses. This research has highlighted how the combined advantages of 6G characteristics, including minimal latency, terahertz communication, and AI-improved threat detection, are transforming the deployment of UAVs. The use of cutting-edge technologies such as blockchain, federated learning, and intelligent reflecting surfaces has enhanced the resilience of drone networks against cyber threats. Even with these progressions, obstacles in scalability, real-time optimization, and smooth integration with current frameworks still pose major challenges. This research lays down a basic structure to address these challenges, offering a new plan to utilize 6G for secure, adaptable, and efficient drone missions in civilian and military fields. Future studies should focus on resolving the current issues with the scalability and energy efficiency of 6G-enabled drones. Investigating the combination of quantum cryptography, improved machine learning techniques, and zero-trust architectures can strengthen system security and scalability. Creating strong datasets for simulating potential threat situations, enhancing resource allocation algorithms for extensive UAV deployments, and examining regulatory frameworks for worldwide standardization are essential for progress.

References

1. Mishra, D., Vegni, A.M., Loscri, V., Natalizio, E.: Drone networking in the 6G era: a technology overview. IEEE Commun. Stand. Maga. **5**(4), 88–95 (2021). https://doi.org/10.1109/mcomstd.0001.2100016
2. Tandra, N., Babu, G., Dhanke, J., Sudhakar, V., Kameswara Rao, M., Ravichandran, S.: Enhancing security and privacy in small drone networks using 6G-IOT driven cyber physical system. Wireless Pers. Commun. (2024). https://doi.org/10.1007/s11277-02411138-8
3. Abdel Hakeem, S.A., Hussein, H.H., Kim, H.: Security requirements and challenges of 6G technologies and applications. Sensors **22**(5), 1969 (2022). https://doi.org/10.3390/s22051969
4. Al-Turjman, F.: Unmanned aerial vehicles in smart cities, 1st edn. Springer, Heidelberg (2020). ISBN: 3030387119
5. Deebak, B.D., Hwang, S.O.: Intelligent droneassisted robust lightweight multi-factor authentication for military zone surveillance in the 6G era. Comput. Netw. **225**, 109664 (2023)
6. Angjo, J., Shayea, I., Ergen, M., Mohamad, H., Alhammadi, A., Daradkeh, Y.I.: Handover management of drones in future mobile networks: 6G technologies. IEEE Access **9**, 12803–12823 (2021)
7. Shvetsov, A.V., et al.: Federated learning meets intelligence reflection surface in drones for enabling 6G networks: challenges and opportunities. IEEE Access **11**, 130860–130887 (2023). https://doi.org/10.1109/access.2023.3323399
8. Amponis, G., et al.: Drones in B5G/6G networks as flying base stations. Drones **6**(2), 39 (2022)
9. Abu, M., Chowdhury, M.Z., Jang, Y.M.: Software-defined uav networks for 6g systems: requirements, opportunities, emerging techniques, challenges, and research directions. IEEE Open J. Commun. Soc. **4**, 2487–2547 (2023)
10. Zahid, M.U., Nisar, M.D., Fazil, A., Ryu, J., Shah, M.H.: Composite ensemble learning framework for passive drone radio frequency fingerprinting in sixth-generation networks. Sensors **24**(17), 5618–5618 (2024)
11. Mahmoud, H.H.H., Amer, A.A., Ismail, T.: 6G: A comprehensive survey on technologies, applications, challenges, and research problems. Trans. Emerg. Telecommun. Technol. (2021). https://doi.org/10.1002/ett.4233
12. Stavroulaki, V., et al.: DEDICAT 6G - dynamic coverage extension and distributed intelligence for human centric applications with assured security, privacy and trust: from 5G to 6G. In: European Conference on Networks and Communications (2021). https://doi.org/10.1109/eucnc/6gsummit51104.2021.9482611
13. Chang, S.-Y., Park, K., Kim, J., Kim, J.: Securing UAV flying base station for mobile networking: a review. Future Internet **15**(5), 176 (2023). https://doi.org/10.3390/fi15050176
14. Song, S., et al.: Analysis of wireless backhaul networks based on aerial platform technology for 6G systems. Comput. Mater. Continua **62**(2), 473–494 (2020). https://doi.org/10.32604/cmc.2020.09052

15. Saad, W., Bennis, M., Chen, M.: A vision of 6G wireless systems: applications, trends, technologies, and open research problems. IEEE Netw. **34**(3), 1–9 (2019)
16. Bandi, A., Yalamarthi, S.: Towards artificial intelligence empowered security and privacy issues in 6G communications. In: 2022 International Conference on Sustainable Computing and Data Communication Systems (ICSCDS) (2022). https://doi.org/10.1109/icscds53736. 2022.9760857
17. Puspitasari, A.A., An, T.T., Alsharif, M.H., Lee, B.M.: Emerging technologies for 6G communication networks: machine learning approaches. Sensors **23**(18), 7709 (2023). https://doi. org/10.3390/s23187709
18. Uwaoma, C.: On Security Strategies for Addressing Potential Vulnerabilities in 6G Technologies Deployable in Healthcare (2023). ArXiv.org. https://arxiv.org/abs/2309.16714
19. Kishk, M., Bader, A., Alouini, M.-S.: Aerial base station deployment in 6G cellular networks using tethered drones: the mobility and endurance tradeoff. IEEE Veh. Technol. Mag. **15**(4), 103–111 (2020). https://doi.org/10.1109/mvt.2020.3017885
20. Akinbi, A.: Digital forensics challenges and readiness for 6G Internet of Things (IoT) networks (2023)
21. Ning, Z., et al.: Intelligent-reflecting-surface-assisted UAV communications for 6G networks (2023). https://arxiv.org/abs/2310.20242
22. Wang, X., Guo, Y., Gao, Y.: Unmanned autonomous intelligent system in 6G non-terrestrial network Information **15**(1), 38 (2024). https://doi.org/10.3390/info15010038
23. Masaracchia, A., et al.: UAV Enabled ultra-reliable low-latency communications for 6G: a comprehensive survey. IEEE Access **9**, 137338–137352 (2021). https://doi.org/10.1109/acc ess.2021.3117902
24. Wang, C.-X., Huang, J., Wang, H., Gao, X., You, X., Hao, Y.: 6G wireless channel measurements and models: trends and challenges. IEEE Veh. Technol. Mag. **15**(4), 22–32 (2020). https://doi.org/10.1109/mvt.2020.3018436
25. Deebak, B.D., Al-Turjman, F.: Drone of IoT in 6G wireless communications: technology, challenges, and future aspects, pp. 153–165 (2020). https://doi.org/10.1007/978-3-030-387 12-9_9
26. Hatami, M., Qu, Q., Chen, Y., Kholidy, H., Blasch, E., Ardiles-Cruz, E.: A survey of the real-time metaverse: challenges and opportunities. Future Internet **16**(10), 379 (2024)
27. Kholidy, H.A.: Dynamic Network Slicing Orchestration in Open 5G Networks using Multi-Criteria Decision Making and Secure Federated Learning Techniques, 08 August 2024, PREPRINT (Version 1). https://doi.org/10.21203/rs.3.rs-4745968/v1
28. Mustafa, F.M., Kholidy, H.A., Sayed, A.F., et al.: Optical fiber fronthaul segment in open radio access 5G networks: enhanced performance utilizing AFBG. Opt. Quant. Electron. **56**, 1014 (2024)
29. Kholidy, H.A., Berrouachedi, A., Benkhelifa, E., Jaziri, R.: Enhancing security in 5G networks: a hybrid machine learning approach for attack classification. In: 2023 20th ACS/IEEE International Conference on Computer Systems and Applications (AICCSA), Giza, Egypt, pp. 1–8 (2023)
30. Almazyad, I., Shao, S., Hariri, S., Kholidy, H.A.: Anomaly behavior analysis of smart water treatment facility service: design, analysis, and evaluation. In: 2023 20th ACS/IEEE International Conference on Computer Systems and Applications (AICCSA), Giza, Egypt, pp. 1–7 (2023)
31. Kholidy, H.A., et al.: Secure the 5G and beyond networks with zero trust and access control systems for cloud native architectures. In: 2023 20th ACS/IEEE International Conference on Computer Systems and Applications (AICCSA), Giza, Egypt, pp. 1–8 (2023). https://doi.org/ 10.1109/AICCSA59173.2023.10479308

32. Boualem, A., Berrouachedi, A., Ayaida, M., Kholidy, H., Benkhelifa, E.: A new hybrid cipher based on prime numbers generation complexity: application in securing 5G networks. In: 2023 20th ACS/IEEE International Conference on Computer Systems and Applications (AICCSA), Giza, Egypt, pp. 1–8 (2023). https://doi.org/10.1109/AICCSA59173.2023.10479316

33. Abushgra, A.A., Kholidy, H.A., Berrouachedi, A., Jaziri, R.: Innovative routing solutions: centralized hypercube routing among multiple clusters in 5G networks. In: 2023 20th ACS/IEEE International Conference on Computer Systems and Applications, Giza, Egypt, pp. 1–7 (2023)

34. Kholidy, H.A.: A smart network slicing provisioning framework for 5G-based IoT networks. In: 2023 10th International Conference on Internet of Things: Systems, Management and Security (IOTSMS), San Antonio, TX, USA, pp. 104–110 (2023). https://doi.org/10.1109/IOTSMS59855.2023.10325712

35. Khalil, A.A., Rahman, M.A., Kholidy, H.A.: FAKEY: fake hashed key attack on payment channel networks. In: 2023 IEEE Conference on Communications and Network Security (CNS), Orlando, FL, USA, pp. 1–9 (2023). https://doi.org/10.1109/CNS59707.2023.10288911

36. Kholidy, H.A., Baiardi, F., Azab, A.: A data-driven semi-global alignment technique for masquerade detection in stand-alone and cloud computing systems. In: Submitted in Granted on January 2019, US 20170019419 A1 (2019)

37. Kholidy, H.A.: Accelerating stream cipher operations using single and grid systems. US Patent and Trademark Office (USPTO), April 2012, US 20120089829 A1 (2019)

38. Kholidy, H.: Multi-layer attack graph analysis in the 5G edge network using a dynamic hexagonal fuzzy method. Sensors **22**, 9 (2022). https://doi.org/10.3390/s22010009.(IF:3.576)

39. Kholidy, H.: Detecting impersonation attacks in cloud computing environments using a centric user profiling approach. Future Gener. Comput. Syst. **117**(17), 299–320 (2021), ISSN 0167–739X. https://doi.org/10.1016/j.future.2020.12.009. https://www.sciencedirect.com/science/article/pii/S0167739X20330715

40. Kholidy, H.: Autonomous mitigation of cyber risks in cyber-physical systems. Future Gener. Comput. Syst. **115**, 171–187 (2021). ISSN 0167–739X

41. Kholidy, H.A.: An intelligent swarm based prediction approach for predicting cloud computing user resource needs. Comput. Commun. J. (2020). https://authors.elsevier.com/tracking/article/details.do?aid=6085&jid=COMCOM&surname=Kholidy

42. Kholidy, H.A.: Correlation based sequence alignment models for detecting masquerades in cloud computing. IET Inf. Secur. J. (2019). https://doi.org/10.1049/iet-ifs.2019.0409. https://digital-library.theiet.org/content/journals/https://doi.org/10.1049/iet-ifs.2019.0409

43. Alkhowaiter, M., et al.: Adversarial-aware deep learning system based on a secondary classical machine learning verification approach. Sensors **23**, 6287 (2023). https://doi.org/10.3390/s23146287

44. Jakaria, A., et al.: Trajectory synthesis for a UAV swarm based on resilient data collection objectives. IEEE Trans. Netw. Serv. Manag. (2022)

45. Kholidy, H.A., Hassan, H., Sarhan, A., Erradi, A., Abdelwahed, S.: QoS optimization for cloud service composition based on economic model. In: The Internet of Things. User-Centric IoT, vol. 150 (2015). ISBN: 978-3-319-19655-8

46. Rahman, A., Mahmud, M., Iqbal, T., Kholidy, H., Saraireh, L., et al.: Network anomaly detection in 5G networks. Math. Model. Eng. Prob. J. 9(2), 397–404 (2022). https://doi.org/10.18280/mmep.090213

47. Kholidy, H.A., et al.: A survey study for the 5G emerging technologies. Acta Scientific Comput. Sci. 5(4), 63–70 (2023). https://doi.org/10.13140/RG.2.2.22308.04485

48. Kholidy, H.A., et al.: A hierarchical cloud intrusion detection system: design and evaluation. Int. J. Cloud Comput. Serv. Arch. (IJCCSA) (2012)

49. Bohn, R., et al.: NIST multi-domain knowledge planes for service federation for 5G & beyond public working group: applications to federated autonomic/autonomous networking. In: The IEEE Future Networks World Forum (FNWF), Baltimore, MD, USA, 13–15 November 2023 (2023)
50. Kholidy, H.A., Karam, A., Sidoran, J., et al.: Toward zero trust security in 5g open architecture network slices. In: IEEE Military Conference (MILCOM), CA, USA, 29 November 2022 (2022). https://edas.info/web/milcom2022/program.html
51. Kholidy, H.A., Reed, K.J.H., Elazzazi, Y.: An experimental 5G testbed for secure network slicing evaluation. In: The 2022 IEEE Future Networks World Forum (FNWF), Montreal, Canada (2022). https://fnwf.ieee.org/wp-content/uploads/sites/339/2022/10/AcceptedPaperScheduleV0.1.pdf
52. Kholidy, H.A., Hariri, S.: Toward an experimental federated 6G testbed: a federated leaning approach. In: The 19th ACS/IEEE International Conference on Computer Systems and Applications (AICCSA 2022), Abu Dhabi, UAE, 5–7 December 2022 (2022)
53. Kholidy, H., Karam, A., Sidoran, J.L., Rahman, M.A.: 5G core security in edge networks: a vulnerability assessment approach. In: The 26th IEEE Symposium on Computers and Communications (The 26th IEEE ISCC), Athens, Greece, 5–8 September 2021 (2021)
54. Haque, N.I., Rahman, M.A., Chen, D., Kholidy, H.: BIoTA: control-aware attack analytics for building internet of things. In: 2021 18th Annual IEEE International Conference on Sensing, Communication, and Networking (IEEE SECON), pp. 1–9 (2021)
55. Iannucci, S., Kholidy, H.A., Ghimire, A.D., Jia, R., Abdelwahed, R., Banicescu, I.: A comparison of graph-based synthetic data generators for benchmarking next-generation intrusion detection systems. In: IEEE Cluster, Hawaii, USA, 5 September 2017 (2017)
56. Chen, Q., Kholidy, H.A., Abdelwahed, S., Hamilton, J.: Towards realizing a distributed event and intrusion detection system. In: The International Conference on Future Network Systems and Security (FNSS 2017), Gainesville, Florida, USA, 31 August 2017 (2017)
57. Kholidy, H.A., Erradi, A.: A cost-aware model for risk mitigation in cloud computing systems. In: 12th ACS/IEEE International Conference on Computer Systems and Applications (AICCSA), Marrakech, Morocco (2015)
58. Kholidy, H.A., Erradi, A., Abdelwahed, S.: Attack prediction models for cloud intrusion detection systems. In: The International Conference on Artificial Intelligence, Modelling and Simulation (AIMS2014), Madrid, Spain (2014)
59. Kholidy, H.A., Yousouf, A.M., Erradi, A., Ali, H., Abdelwahed, S.: A finite context intrusion prediction model for cloud systems with a probabilistic suffix tree. In: The 8th European Modelling Sympos on Mathematical Modelling and Comp Simulation, Pisa, Italy (2014)
60. Kholidy, H.A., Erradi, A., Abdelwahed, S.: Online risk assessment and prediction models for autonomic cloud intrusion prevention systems. In: The "11th ACS/IEEE International Conference on Computer Systems and Applications (AICCSA), Doha, Qatar (2014)
61. Kholidy, H.A., Erradi, A., Abdelwahed, S., Azab, A.: A finite state hidden markov model for predicting multistage attacks in cloud systems. In: The 12th IEEE International Conference on Dependable, Autonomic and Secure Computing (DASC), Dalian, China (2014)
62. Kholidy, H.A., Erradi, A., Abdelwahed, S., Baiardi, F.: A hierarchical, autonomous, and forecasting cloud IDS. In: The 5th International Conference on Modeling, Identification and Control (ICMIC2013), Cairo, 31 August–2 September 2013 (2013)
63. Arshad, M., Tirth, P., Kholidy, H.: Deception technology: a method to reduce the attack exposure time of a SCADA System. https://dspace.sunyconnect.suny.edu/handle/1951/70148,
64. Bhoite, A., Basnet, D., Kholidy, H.: Risk Evaluation for Campus Area Network. https://dspace.sunyconnect.suny.edu/handle/1951/70162
65. Malkoc, M., Kholidy, H.A.: 5G Network Slicing: Analysis of Multiple Machine Learning Classifiers (2023). ArXiv https://arxiv.org/abs/2310.01747

66. Kholidy, H.A., Abuzamak, M.: 5G network management, orchestration, and architecture: a practical study of the MonB5G project (2022). https://doi.org/10.48550/arXiv.2212.13747
67. Abuzamak, M., Kholidy, H.: UAV based 5G network: a practical survey study (2022). https://doi.org/10.48550/arXiv.2212.13329
68. Kholidy, H. A., Rahman, M. A., Karam, A., Akhtar, Z.: Secure spectrum and resource sharing for 5G Networks using a blockchain-based decentralized trusted computing platform (2022). https://doi.org/10.48550/arXiv.2201.00484
69. Kholidy, H.A.: A triangular fuzzy based multicriteria decision making approach for assessing security risks in 5G networks (2021). https://doi.org/10.48550/arXiv.2112.13072
70. Kholidy, H.A., Balbuena, W., Mustafa, F.M.: A survey study for the 5G emerging technologies (2023). https://doi.org/10.13140/RG.2.2.22308.04485
71. Fox, A., Kholidy, H.A., Almazyad, I.: Current 5G federation trends: a literature review (2023). https://doi.org/10.13140/RG.2.2.33470.46405
72. Kholidy, H.A., Anjony, T., Almazyad, I.: A survey of the AI and cybersecurity solutions used for network slicing in the 5G and beyond domain (2023)

Unlocking the Potential of Containers in Scientific Computing to Achieve Bitwise Reproducibility, Portability and Performance

Jean Iaquinta[1]([✉]) [ID] and Anne Fouilloux[2] [ID]

[1] Norwegian Research Infrastructure Services, Oslo, Norway
jeani@uio.no
[2] Simula Research Laboratory, Oslo, Norway
annef@simula.no

Abstract. The modern form of containers, as popularly known through platforms like Docker, Singularity (Apptainer), Podman or Charliecloud, to cite only a few, began to take shape over a decade ago. However, the fundamentals behind containerization and actual benefits of containers in scientific computing remain largely unclear to a vast majority of users. In fact, there is a significant gap between simplistic "Hello world" examples found online and real scientific applications. Also, rumors suggest that achieving satisfactory performances on supercomputers across multiple nodes is impossible.

The aim of this paper is to explain how containers can leverage the potential of high-speed networks for inter-node communications with UCX on Fram and Betzy (from the Norwegian national e-infrastructure provider). It is also shown how to achieve near-native performance on LUMI (EuroHPC's flagship) despite a "Slingshot-11" interconnect and proprietary library.

Results obtained in the standard OSU Micro-Benchmarks tests for latency and bandwidth, and with a fully-fledged climate model, demonstrate that containerized applications work just as well as their bare-metal counterparts, are portable and provide bit-for-bit reproducibility on different platforms.

Containers are therefore highly recommended to minimize deployment and porting issues *i)* for AaaS (Applications as a Service) coming with all the necessary software environment (rather than source code only); and *ii)* so that HPC users do not have to rely on anybody to install what they need and can be operational within minutes whilst still getting top performance.

Keywords: Containers · inter-node communication · reproducibility

1 Introduction

A container, as defined by Docker, Inc., is a *"standard unit of software that packages up code and all its dependencies so the application runs quickly and reliably*

© The Author(s) 2025
A. Azab and T. Malkiewicz (Eds.): NeIC 2024, CCIS 2398, pp. 45–60, 2025.
https://doi.org/10.1007/978-3-031-86240-3_4

from one computing environment to another" (https://www.docker.com). Containerization is not a new concept, but it is often still considered as a "toy" rather than a necessity, especially on supercomputers where containers are largely discarded in favour of specialized package managers like EasyBuild [6] or Spack [4] to install software and resolve dependencies. Indeed, the fundamentals behind containerization as well as benefits containers can bring to scientific computing remain unclear to a vast majority of users, and also in the HPC (High Performance Computer) community itself.

So what primarily prevents a wider adoption of containers for scientific applications in general, and in climate modeling in particular?

Above all there are psychological barriers as well as misunderstandings, a lack of expert knowledge, an absence of efficient and reproducible cases for container usages, as well as persistent rumours about poor performances. There are also technical difficulties due to the complexity of HPC systems' hardware and firmware, to the non-openness of source code and in some instances restrictions on usage/distribution of proprietary compilers and libraries, as is the case for LUMI (https://www.lumi-supercomputer.eu/eurohpcju) with HPE's (Hewlett Packard Enterprise) "Cassini" network interface.

This paper aims at overcoming such prejudices by only exploiting open source compilers and libraries to provide examples of outstanding results showing that containerized applications generally match and often outperform their "bare-metal" counterparts. Containers can easily be transferred and executed on various platforms where they provide a consistent software environment and bit-for-bit reproducibility (for thread-safe deterministic algorithms), thereby representing a step further towards Open Science and Reproducible Research.

To improve the FAIRness of all the research artifacts pertaining to this work, we gathered them together into a so-called "Research Object" on RoHub (the Research Object Hub) [3]. This Research Object [2] aggregates all the scripts, Dockerfiles (i.e., recipes or definitions), container images (stored as Singularity Image Files), relevant results, links to references and useful material, etc.

2 Background

Numerous scientific applications take advantage of parallelism with "MPI" (Message Passing Interface) to concurrently execute tasks or processes and hence speed-up the resolution of complex numerical problems. This often involves distributing the workload among several CPU (Central Processing Unit) "cores" that exchange data with each other and are physically located in different systems or "nodes" in a cluster. On a HPC the various nodes are connected through a high-bandwidth low-latency network or "interconnect".

2.1 Description of the Hardware and Software

To prototype container recipes and experiment with different strategies for production workloads we had access to Fram and Betzy HPC clusters from Sigma2, the Norwegian national infrastructure (https://documentation.sigma2.no/), and LUMI at CSC, the Center of Scientific Computing in Finland (https://docs.lumi-supercomputer.eu):

- Fram is a Lenovo NeXtScale NX-360 with Intel Broadwell 2.10 GHz processors and an InfiniBand network at 100Gbits/s organized in an "island" topology;
- Betzy is a BullSequana XH-2000 with AMD Epyc 7742 2.25 GHz ("Rome") processors and a Mellanox InfiniBand HDR-100 network also providing a 100 Gbits/s data rate;
- LUMI is an HPE Cray-EX whose CPU-partition consists of AMD Epyc 7763 2.45 GHz ("Milan") processors connected by 200 Gbits/s HPE Slingshot-11 (https://www.hpe.com/no/en/compute/hpc/slingshot-interconnect.html).

On Fram and Betzy both Intel MPI and OpenMPI are supported, whereas on LUMI it is only Cray-MPIch vendor-specific versions. There are other differences between these machines, in particular Fram runs Rocky Linux 9.2, Betzy an operating system based on RHEL (Red Hat Enterprise Linux) 7.7, and LUMI SLES (Suse Linux Enterprise Server) 15 Service Pack 4. Fortunately all these HPCs make Singularity or Apptainer available for users to run containers without root privileges, provided they can bring their own .sif (Singularity Image files) either built directly from a .def (Definition File) or after conversion from a Docker archive.

2.2   MPI Applications

Installing any parallel computing application within a container is always a challenge. In the following sections, several container recipes are presented as well as the results of tests for point-to-point latency and one-sided bandwidth carried out with the OSU (Ohio State University) Micro-Benchmarks version 7.3 (https://mvapich.cse.ohio-state.edu/benchmarks) with containers and an MPI library directly installed on the host system. Although there is no guaranty that a container performing well in the benchmark is going to do so for a real-world application, failure in these tests is a serious indication that there is something wrong with inter-nodes communications.

Simple Container. A straightforward option for users who are not system engineers, administrators or HPC specialists is to select an operating system, compiler and MPI version they are comfortable with. They may know their core application (in this case the OSU Micro-Benchmarks) but not much about how to install other dependencies and hence generally leave default configuration parameters, for example following the recipe "Dockerfile_ubuntu2204_mpich412_osu73" stored on RoHub.

The actual built container image "ubuntu2204_mpich412_osu73.sif" can also be found in the same Research Object, as well as the Slurm script prepared to submit the job on Fram for two CPUs belonging to the same node and to two different nodes ("job_1node_simple.sh" and "job_2nodes_simple.sh", respectively).

Figure 1 shows the results of the OSU Micro-Benchmark get bandwidth test and Fig. 2 that for the latency test for the simple container *versus* the "bare-metal" equivalent with OpenMPI4.1.5 executed with CPUs on the same node (to evaluate intra-node communications) or on different nodes (to assess inter-node communications).

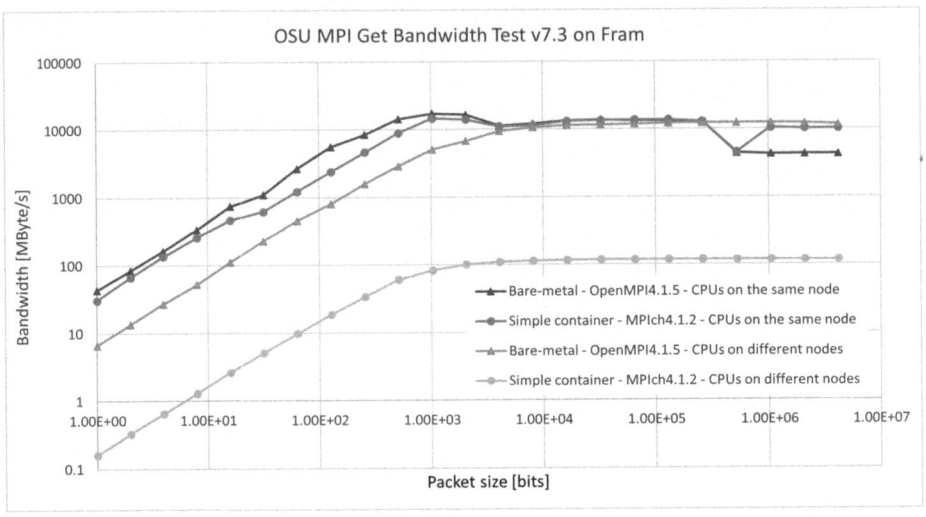

Fig. 1. OSU bandwidth test on Fram with MPIch4.1.2 in the simple container and the "bare-metal" OpenMPI4.1.5 for CPUs on the same node or on different nodes (notice that the plot uses a log-log scale).

The simple container provides similar intra-node values (i.e., between 2 CPUs belonging to the same node) for the bandwidth and latency as the bare-metal. However when it comes to CPUs on different nodes, the latency (or delay in data communications between two processes) is about 60 times larger for the simple container than the bare-metal. The bandwidth also (that is the amount of data which can be retrieved from a remote memory location per unit of time) is on the average 70 times smaller for the container. This is because packets can only resort to the TCP protocol instead of using the high-speed network since the container is lacking appropriate libraries.

Adding a Communication Layer. Most recently released MPI libraries incorporate a device layer implementation for network communication using UCX [10] (Unified Communication X) able to exploit high-speed networks for inter-node communications and shared memory mechanisms for intra-node communication. UCX provides an abstraction layer that allows the same application to run on different types of interconnect technologies including InfiniBand, Ethernet, and RoCE (RDMA over Converged Ethernet) among others, thereby ensuring portability and reducing the need for duplicating efforts for different software stacks.

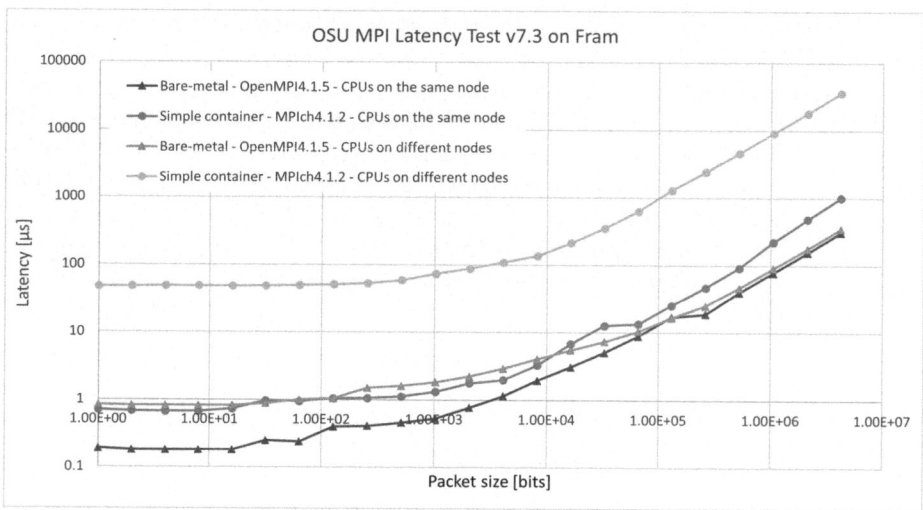

Fig. 2. Similar as Fig. 1 but for the OSU Latency test (also in a log-log scale).

UCX is installed on Fram and Betzy, and it was also available on LUMI until a recent upgrade but it is not supported there any longer (as discussed in more details in the next section); It is precisely something like this UCX communication layer that is missing in the simple container definition.

MPIch versions posterior to 3.3.1 include a new device layer implementation named "CH4" with UCX support for RDMA (Remote Direct Memory Access), enabling fast and hardware-offloaded data transfer. Building on the simple container definition, the "communication layer aware" container recipe was developed using MPIch with UCX support. The most important change is the use of the "CH4=UCX" configure option in the recipe and UCX with its dependencies.

Such recipes were extensively tested on a wide range of HPCs including Fram and Betzy where they provided excellent performance. Note that the principle can easily be extended to other MPI implementations like Open-MPI, MVAPICH2 or Intel MPI and is absolutely not restricted to MPIch. Also it is quite simple to start the build from a different operating system than Ubuntu:22.04 and adapt the recipe accordingly, since apart from a few basic packages everything else is installed from source (see "Dockerfile_osuse153_gcc122_ucx115_mpich412_osu73" on RoHub).

The corresponding Slurm job script is nearly the same as those for the simple container, only the name of the image changes, and the command to submit the parallel tasks is also identical, so that only the libraries and packages installed inside the container are used, therefore we refer to these runs as "standalone".

Although it is often mentioned that an MPI implementation inside a container has to be "compatible" with that on the host, one can notice here that it is not made use of "mpirun" or "mpiexec" to submit the job (which would have required to load an MPI module and raised these compatibility issues) but of the

Slurm command "srun" instead. This is because through MPIch the container supports PMI-2 (the second version of the Process Management Interface) that deals with the initialization, management, and coordination of parallel processes.

The graph on Fig. 3 presents the OSU bandwidth tests for a container built from openSUSE/leap:15.3 with MPIch4.1.2 and UCX1.15.0 *versus* bare-metal OpenMPI4.1.5 for CPUs belonging to two different nodes on Fram and Betzy. Clearly, the standalone container nearly matches the native performance over the full range of packet sizes, approaching the theoretical maximum bandwidth (i.e., $100/8 = 12.5$ GByte/s) for the largest message sizes, and this is already quite an impressive achievement.

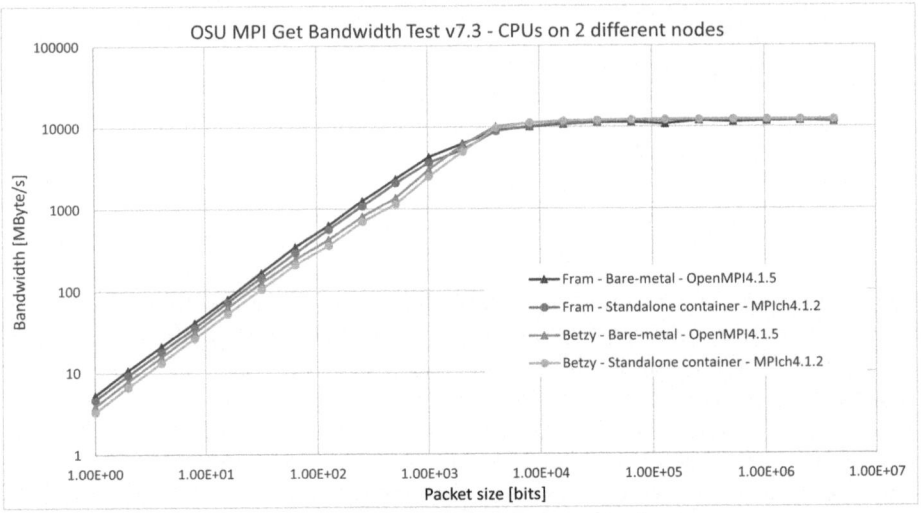

Fig. 3. OSU bandwidth test on Fram and Betzy with a container featuring MPIch4.1.2 and UCX and the equivalent "bare-metal" OpenMPI4.1.5 for CPUs on different nodes (log-log scale).

Using System Libraries from the Host. On LUMI the Slingshot-10 network installed originally, and which at the time supported UCX, was recently replaced by a Slingshot-11 interconnect and HPE "Cassini" network interface cards along with their proprietary libraries. Because neither the sources for Cray-MPIch nor for the "CXI" OFI (OpenFabrics Interfaces) are open [9] it is impossible to build standalone containers for use on multiple nodes on LUMI, and a workaround was needed.

One option to still be able to run containers in these conditions would be to literally "bind" the MPI installed inside the container to that available on the host system as done in [8]. However replacing the container MPI changes the numerical results and ruins portability efforts, therefore it is not the most suited method, especially for reproducibility across different hosts.

Another unconventional way to execute containers is to use high-level APIs (Application Programming Interfaces) available internally, so that the numerical results are not modified, along with cherry-picked low-level libraries from the host itself. The rational is that such libraries being typically related to the hardware and firmware were *a priori* better tuned for a particular supercomputer by its vendor and administrators than what can be done in a container.

The idea, sparkled by [7], revolves around building a container with MPIch support for a "dummy" Libfabric (see it as a placeholder) and swap it at runtime for that on the host, intervening in effect one step before the MPI binding of [8]. An advantage of this method is that it does not require to install all the dependencies for Libfabric, as was the case for UCX, which greatly simplifies the recipe. However, for that to work at all, the container has to be built from versions of the operating system and of the OFI close enough to those on the host. Since no Suse Linux Enterprise Server base image was available it is openSUSE/leap:15.3 that was chosen for "Dockerfile_osuse153_gcc122_ofi114_mpich412_osu73".

Basically the libraries are to be searched in this order: the Libfabric from the host has to come first, then all the libraries inside the container, finally anything still missing is taken from system folders on the host. The Slurm script to submit the job on LUMI with the appropriate bindings and environmental variables is available in the Research Object ("job_2nodes_ofi_swap.sh").

On Fig. 4 is represented the result from the OSU bandwidth test performed on LUMI with the bare-metal CrayMPIch3.4a2 *versus* the MPIch4.1.2 container and OFI swap with the host for CPUs on different nodes. Note that here the curves overlap so perfectly that one of the lines (in this instance that corresponding to the container) had to be made thicker and transparent for the sake of visibility.

2.3   Real-World Application and Reproducibility

Rationale. In the previous sections it was established that containers can provide near-native performance over multiple nodes in standard benchmark tests whether they are used in a standalone mode (as on Fram and Betzy) or whilst sharing low-level libraries from the host (on LUMI). Two important remaining questions are: how do containers perform in "sheer scientific computing", and do they still maintain bit-for-bit reproducibility on machines with x86-64 or amd64 processor architectures?

To address these questions the selected application is the latest version of a state-of-the-art ESM (Earth System Model) called CESM2.3α17a (e.g., the Community Earth System Model) coming with a spectral elements grid based on a continuous Galerkin spectral finite-element method applied on a "cubed-sphere" [11]. One of the benefits of the spectral elements dynamical core is that the results of the simulation does not depend on the number of processors used, whereas with the finite volumes that would have modified the domain decomposition, also the former provides a much better scalability.

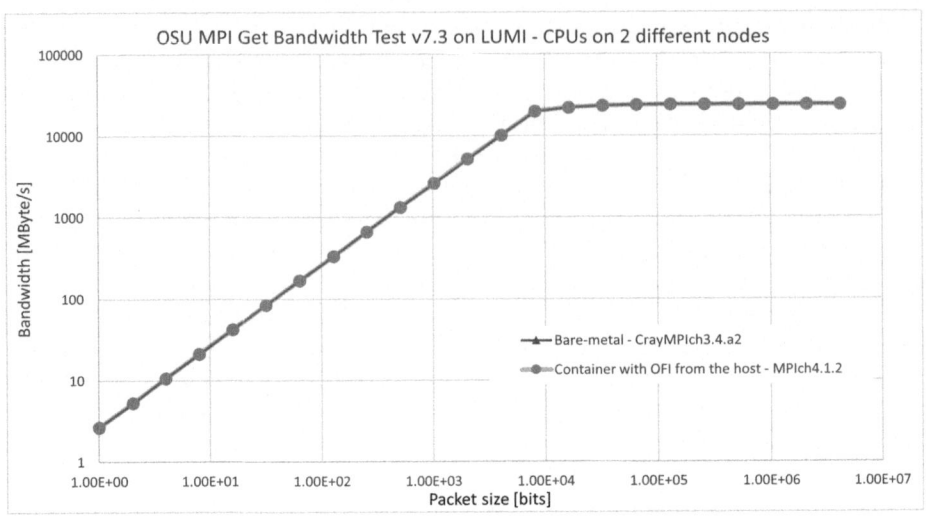

Fig. 4. OSU bandwidth test on LUMI with a container featuring MPIch4.1.2 and OFI swap with the host and the equivalent "bare-metal" CrayMPIch3.4a2 for CPUs on different nodes (log-log scale).

A new case was created with CESM2.3α17a featuring the classic *F2000climo* components set with an *ARCTICGRIS$\times$8* grid. This configuration is characterized by a notoriously demanding variable resolution mesh with 152390 vertical columns and nested grids centered on the North Pole (where the horizontal resolution is progressively increased by a factor of $\times$4, that is equivalent to about about 25 km from a global grid corresponding to about 100 km for the remainder of the globe, plus a further x2 local refinement broadly covering Greenland and the Iceland, hence $\times$8 in total).

The duration of the simulation was classically set to 20 days and the job was submitted on LUMI, with an openSUSE/leap:15.3 MPIch4.1.2 container swapping OFI with the host (see "job_esm_ofi_swap.sh"), then on Fram and Betzy with a standalone container based on the same operating system and MPI but supporting UCX like that used for Fig. 3. An example of what the model can produce is also plotted on Fig. 5.

Containers for Earth System Modelling. There are lots of different aspects to consider when assessing the performance of the containers in this modelling exercise. One possibility is to compare the actual run time for the simulations as a function of the number of nodes or CPUs used (1 node on Betzy or LUMI is composed of 128 CPUs, 1 node on Fram has 32 CPUs), as reported in Table 1, but such numbers do not mean much for climate scientists. The numerical performance of an Earth System Model is more commonly expressed in terms of

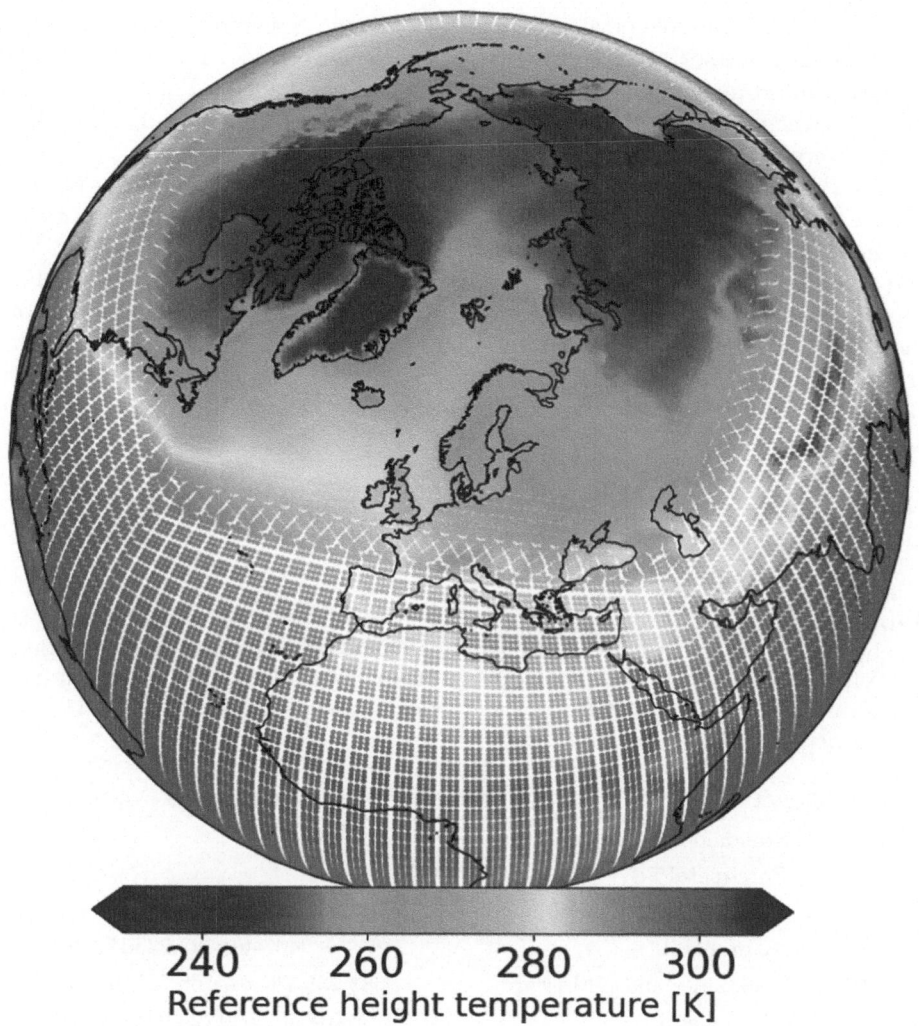

Fig. 5. Example of monthly mean reference height (i.e., at 2 m) air temperature (TRE-FHT) expressed in kelvin obtained with CESM2.3α17a and corresponding to a typical situation in January 2000, starting from a global $1° \times 1°$ grid which is progressively increased by a factor of $\times 4$ over the Arctic, and exhibits a further $\times 2$ resolution refinement over Greenland and Iceland.

model throughput, that is to say the number of simulated climate years per computational day for a given amount of computing resources. The outcomes of this comparison are summarized in Table 2 which gives the model throughput as a function of the number of nodes used.

Note that for the sole purpose of checking the reproducibility between x86-64 and amd64 processor architectures, and also because it is a much smaller and heavily utilized machine, only a few simulations were made on Fram. Anyway, it was sufficient to confirm that the results obtained on x86-64 or amd64 processor architectures are bit-for-bit identical.

For the sake of comparison between container *versus* bare-metal, a number of runs were submitted using packages "equivalent" to those inside the container but available as modules directly installed on Fram and Betzy, that is GNU11.3 and OpenMPI4.1.4. It has to be mentioned that ESMF, the Earth System Modeling Framework (https://github.com/esmf-org/esmf), which was absent from the LUMI software stack was built manually in the same way as in the containers and with the same options. This is one more reason for using containers: they can be up and running very quickly, regardless of the software available on the host.

The corresponding values are included in Tables 1 and 2, and the throughput plotted as well on Fig. 6: clearly the containers largely outperform their bare-metal counterpart, on all the HPCs.

Table 1. CESM2.3α17a average run time in seconds/day as a function of the number of CPUs for openSUSE/Leap15.4 - GNU12.2 containers on Fram/Betzy (UCX1.15 - MPIch4.1.2) and LUMI (OFI1.14 swap - MPIch4.1.2), as well as bare-metal runs on Fram/Betzy (GNU11.3 and OpenMPI4.1.4) and LUMI (GNU11.2 and CrayMPIch3.42a2). The lower the better.

Number of CPUs	256	512	1024	2048	4096
Fram - Standalone UCX container	1480.237	810.563			
Fram - Bare-metal	1597.453	890.914			
Betzy - Standalone UCX container		733.999	369.749	193.329	106.660
Betzy - Bare-metal			407.401	222.235	158.447
LUMI - Container and OFI swap		683.238	336.167	166.731	91.993
LUMI - Bare-metal		776.819	402.529	224.056	145.205

An interesting comparison is between simulations carried out on LUMI and on Betzy: the former are on average about 12% faster and hence provide a better model throughput. This is not surprising given the differences in hardware and also the fact that the effective bandwidth offered by the Slingshot-11 on LUMI is twice that of the Mellanox InfiniBand.

Table 2. CESM2.3α17a model throughput in simulated years/day as a function of the number of CPUs for openSUSE/Leap15.4 - GNU12.2 containers on Fram/Betzy (UCX1.15 - MPIch4.1.2) and LUMI (OFI1.14 swap - MPIch4.1.2), as well as bare-metal runs on Fram/Betzy (GNU11.3 and OpenMPI4.1.4) and LUMI (GNU11.2 and CrayMPIch3.42a2). The higher the better.

Number of CPUs	256	512	1024	2048	4096
Fram - Standalone UCX container	0.16	0.29			
Fram - Bare-metal	0.15	0.27			
Betzy - Standalone UCX container		0.32	0.64	1.22	2.28
Betzy - Bare-metal			0.58	1.07	1.49
LUMI - Container and OFI swap		0.35	0.70	1.42	2.57
LUMI - Bare-metal		0.30	0.59	1.06	1.63

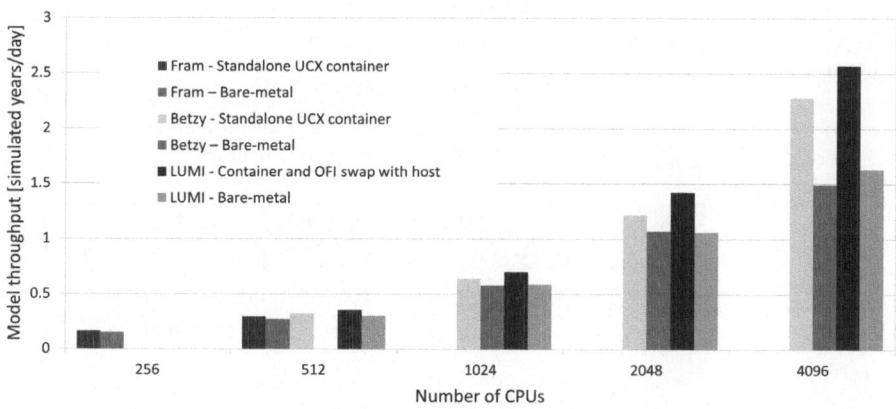

Fig. 6. Model throughput achieved with CESM2.3α17a as a function of the number of CPUs used, for bare-metal and container simulations carried out on Fram, Betzy and LUMI. The higher the better.

When it comes to reproducibility of climate modelling experiments, it is very convenient to monitor four integrated global outputs, either during the run or at the end of the simulations, namely:

- E_{inp}^{glob} the global mean energy of input state [J];
- E_{out}^{glob} the global mean energy of output state [J];
- H^{glob} the global mean heating rate [J/s];
- P_{surf}^{glob} the global mean surface pressure [Pa].

These numbers are calculated after each time step, from the model state, and can be considered as a "digital fingerprint" uniquely representing a dataset (like

"SHA-512" or cryptographic hashes). They are particularly sensitive and betray any change made to the model source code, to the input data or to the computing environment, thereby offering an easy way to compare two simulations since they will start to deviate after a single time step if something is different.

Table 3. CESM2.3α17a digital fingerprints after 20-day bare-metal simulations carried out on Fram/Betzy using GNU11.3 and OpenMPI4.1.4 from modules available on the host.

Global energy (fingerprints)	Fram Bare metal	Betzy Bare metal
E_{inp}^{glob}	0.26039124214589081E+05	0.26053774723676643E+10
E_{out}^{glob}	0.26039124834378023E+10	0.26053776048519101E+10
H^{glob}	0.27486358652322133E-04	0.58753937035464647E-04
P_{surf}^{glob}	0.98500528278332058E+05	0.98500675248831540E+05

Table 4. CESM2.3α17a digital fingerprints after 20-day simulations carried out with openSUSE/Leap15.4 - GNU12.2 - MPIch4.1.2 containers on Fram/Betzy (standalone UCX1.15) and LUMI (OFI1.14 in the container replaced by OFI1.15.2.0 library from the host).

Global energy (fingerprints)	Fram/Betzy Standalone UCX container	LUMI Container and OFI swap with host
E_{inp}^{glob}	0.26063123734497628E+10	0.26063123734497628E+10
E_{out}^{glob}	0.26063124967423949E+10	0.26063124967423949E+10
H^{glob}	0.54677003192420576E-04	0.54677003192420576E-04
P_{surf}^{glob}	0.98501842179353640E+05	0.98501842179353640E+05

As could be expected, bare-metal simulations from Fram/Betzy are slightly different (see Table 3). However, with containers, the integrated model outputs are absolutely identical after 20 days and 7681 time steps (see in Table 4), despite having been obtained on three different HPCs with two different container images: one executed as standalone with UCX on Intel (Fram) and AMD "Rome" (Betzy) machines, and the other one on an AMD "Milan" (LUMI) machine swapping OFI libraries with the host. Basically adopting different transports does not impact the final numerical results and makes it possible to achieve the best possible performance for different use cases. The different types of processors did not matter either because compilation options likely to be architecture dependent were disabled when building the CESM executable.

3 Discussion

"Reproducibility" is essential in computational research in general and for Earth System Modelling in particular, however scientists often have divergent understandings about what is behind this term and how far they have to go to make their research reproducible [5]. The lack of reproducibility of climate simulations is a long-standing issue, as reminded in [1] *"a CESM simulation output is only bit-reproducible when the exact same code is run using the same CESM version and parameter settings, initial conditions and forcing data, machine, compiler (and flags), MPI library, and processor counts, among others. Unfortunately, control over these quantities to this degree is virtually impossible to attain in practice, and further, because the climate system is nonlinear and chaotic, even a double-precision round off-level change will propagate rapidly and result in output that is no longer bit identical to the original."*

It is now within reach to answer confidently that with containers reproducibility is not "impossible to attain in practice" any more, even when switching from an Intel to an AMD machine, or vice-versa. Furthermore with our standalone UCX based containers one can literally create a new use case with CESM, set it up and compile it on a machine based on processors supporting the IEEE 754 standard for floating-point arithmetic, then seamlessly transfer the executable ("cesm.exe") onto another of these computers and proceed to run it as if all had been done on it.

There are indeed significant benefits in using containers providing a consistent software environment on various computers and getting bit-for-bit identical model outputs. It means that with containers researchers will effectively be able to carry out their ESM simulations wherever computing resources are available, or start them somewhere and continue elsewhere, and still get exactly the same results as if they had not changed. It also ought to increase trustworthiness, and in terms of workflow it greatly simplifies job orchestration since there is no more need to worry about where exactly ESM simulations are executed: no "climate changing" differences from the baseline reference will be introduced.

4   Conclusion

In the present paper several approaches for building and using containers were documented, tested and validated for scientific computing on multiple node HPCs, with a particular emphasis on Earth System Modeling. It was demonstrated that, when built and executed with appropriate options, containers provide excellent performance, bit-for-bit reproducibility (as long as all CPU architecture dependent options are deactivated) and portability.

Making the very same containers available to researchers, for model development or testing on their personal computer or Virtual Machines should be an incentive for them to move part of the workload from HPC towards cheaper and more energy/environmentally friendly computing resources. Running a container on a laptop is also more comfortable and responsive since there is no need to be connected to the Internet, no queuing time, etc. Containers can also serve for training/teaching purposes, and eventually become part of workflows, thereby contributing to foster more collaborations, reproducible research and a wider adoption of Open Science.

In an ideal world, every scientific application released should come with a fully validated container, or better yet, the one used by the developers themselves. Such a practice would virtually eliminate the wasteful expenditure of time and resources required to install these applications on different hosts. There would be no need for users to "reinvent the wheel", that is, to recreate a working software environment from scratch based on their understanding of the requirements and dependencies and on what is available on their machine. It would also erase ambiguities about how applications are run in different environments, reduce the demand for basic user support to help newcomers get up and running, and offer additional advantages in terms of efficiency and ease of use.

In terms of performance, generic containers already provide very satisfactory results, although additional tuning could be explored. The "base" container recipes developed in this work could be further modified, for example, to incorporate different compilers, more advanced mathematical libraries, and other MPI implementations; hence, there is still potential for additional speedup.

Applications like ESMs are notoriously MPI bound, and therefore after a few hundred CPUs the effectiveness of the communications has more impact on performance than core-level optimization like FMA (Fused Multiply-Add) which would introduce discrepancies between runs on different machines. By not enabling result changing options and optimizations it is reproducibility that was *de facto* prioritized, however such flags can be reintroduced at any time if a higher throughput is wanted (especially for small configurations) at the expense of bit-identical outputs though.

In fact reproducibility is one of these things people implicitly take for granted in every day's life: who would be happy if for example the integral content of a text document changed "by itself" from one computer to another, or if spreadsheets gave different results? However, in high-performance computing, scientists had no other choices but to accept that applications can behave somewhat erratically on different computers and/or produce different results. The utilization of containers as presented in this paper could put an end to this surreal situation.

Acknowledgements. The authors are very grateful to Uninett Sigma2 who manages Norway's e-infrastructure for large-scale data and computational sciences and provided HPC access to the supercomputers Fram, Betzy and LUMI (as part of projects NN1000K for computing and NS1000K for storage).

The research presented in this paper also benefited (in particular for building and testing the containers) from the Experimental Infrastructure for Exploration of Exascale Computing (eX3), which is financially supported by the Research Council of Norway under contract 270053.

Funding Information. Initiated under the NeIC (Nordic Collaboration on e-Infrastructures for Earth System Modeling) NICEST2 (https://neic.no/nicest2) project, this work was finalized under the EuroScienceGateway (https://www.euro sciencegateway.org) project. NICEST2 was supported by NordForsk (https://www.nordforsk.org), an organization under the Nordic Council of Ministers who provides funding and facilitates Nordic cooperation on research and research infrastructure.

EuroScienceGateway is receiving funding from European Union programme Horizon Europe (HORIZON-INFRA-2021-EOSC-01-04) under grant agreement number 101057388 and the UK Research and Innovation (UKRI) under UK government's Horizon Europe funding guarantee grant number 10038963.

References

1. Ahn, D.H., et al.: Keeping science on keel when software moves. Commun. ACM **64**(2), 66–74 (2021). https://doi.org/10.1145/3382037
2. Fouilloux, A., Iaquinta, J.: Unveiling the potential of containers in scientific computing (2023). https://w3id.org/ro-id/18d3e6d6-e559-4e43-b464-55275bd14708
3. Fouilloux, A., Trasatti, E., Foglini, F., Coca-Castro, A., Iaquinta, J.: Fair research objects for realising open science with the eosc project reliance. Res. Ideas Outcomes **9** (2023). https://doi.org/10.3897/rio.9.e108765
4. Gamblin, T., et al.: The spack package manager: bringing order to hpc software chaos. In: Proceedings of the International Conference for High Performance Computing, Networking, Storage and Analysis, SC 2015. ACM (2015). https://doi.org/10.1145/2807591.2807623
5. Hill, D.R.: Numerical reproducibility of parallel and distributed stochastic simulation using high-performance computing. In: Traoré, M.K. (ed.) Computational Frameworks, pp. 95–109. Elsevier (2017). https://doi.org/10.1016/B978-1-78548-256-4.50004-1
6. Hoste, K., Timmerman, J., Georges, A., De Weirdt, S.: Easybuild: building software with ease. In: 2012 SC Companion: High Performance Computing, Networking Storage and Analysis, pp. 572–582 (2012). https://doi.org/10.1109/SC.Companion.2012.81
7. Madonna, A., Aliaga, T.: Libfabric-based injection solutions for portable containerized mpi applications. In: 2022 IEEE/ACM 4th International Workshop on Containers and New Orchestration Paradigms for Isolated Environments in HPC (CANOPIE-HPC), pp. 45–56 (2022). https://doi.org/10.1109/CANOPIE-HPC56864.2022.00010
8. Sande Veiga, V., et al.: Evaluation and benchmarking of singularity mpi containers on eu research e-infrastructure. In: 2019 IEEE/ACM International Workshop on Containers and New Orchestration Paradigms for Isolated Environments in HPC (CANOPIE-HPC), pp. 1–10 (2019). https://doi.org/10.1109/CANOPIE-HPC49598.2019.00006
9. Shafie Khorassani, K., Chen, C.C., Ramesh, B., Shafi, A., Subramoni, H., Panda, D.: High performance mpi over the slingshot interconnect: Early experiences. In: Practice and Experience in Advanced Research Computing, PEARC 2022. Association for Computing Machinery, New York (2022). https://doi.org/10.1145/3491418.3530773
10. Shamis, P., et al.: Ucx: an open source framework for hpc network apis and beyond. In: 2015 IEEE 23rd Annual Symposium on High-Performance Interconnects, pp. 40–43 (2015). https://doi.org/10.1109/HOTI.2015.13
11. Zarzycki, C.M., Jablonowski, C., Taylor, M.A.: Using variable-resolution meshes to model tropical cyclones in the community atmosphere model. Month. Weather Rev. **142**(3), 1221–1239 (2014). https://doi.org/10.1175/MWR-D-13-00179.1. https://journals.ametsoc.org/view/journals/mwre/142/3/mwr-d-13-00179.1.xml

Integration of Nordugrid ARC with Galaxy and EGI IM

Maiken Pedersen[1]([✉]), Balazs Konya[2], Sebastian Luna-Valero[3], and Björn Grüning[4]

[1] University of Oslo, Oslo, Norway
maikenp@uio.no
[2] Lund University, Lund, Sweden
balazs.konya@hep.lu.se
[3] EGI Foundation, Amsterdam, Netherlands
sebastian.luna.valero@egi.eu
[4] University of Freiburg, Breisgau, Germany

Abstract. In the scientific domain of high energy physics (HEP), the Worldwide LHC Computing Grid (WLCG) was created in order to handle the huge compute and storage needs of the experiments at Large Hadron Collider (LHC). Today WLCG combines about 1.4 million computer cores and 1.5 exabytes of storage from over 170 sites in 42 countries. What ties the sites together is the middleware installed at each site, one of these being the Nordugrid ARC (Advanced Resource Connector).

ARC has been a great success and has served, and continues to serve the HEP community very well. Up until now though, we have had limited success in sharing our technology with other communities, despite the fact that many are faced with challenges that ARC solves: managing computation and storage across different infrastructure providers.

With a network of ARC enabled compute sites - a user can submit a job from "anywhere" and automatically be routed to the best site depending on various matchmaking rules. One of the key strengths of ARC is its inbuilt data handling capabilities. ARC seamlessly downloads any remote input data to the computing site and makes sure all data is in place before the job is passed to the site's local batch system. Once the job is done ARC can upload the data to a remote storage site, or it can be manually retrieved.

In this paper we describe how we have integrated ARC with the Galaxy Project portal in the context of the EuroScienceGateway project. The Galaxy portal is a user-friendly job-submission and workflow platform that lets a user easily define and submit jobs to an underlying computing cluster, it allows reproducibility in addition to facilitating sharing of jobs and workflows. The Galaxy project has a large user-base from the bioinformatics communities, in addition to users from the climate, astrophysics and material science communities, to mention a few. With ARC integration in Galaxy, these new communities will seamlessly be able to enjoy the benefits of ARC by using Galaxy to submit jobs to their remote HPC system, instead of having to manually log into the HPC system and interact with the local batch system via scripting. We

A. Azab and T. Malkiewicz (Eds.): NeIC 2024, CCIS 2398, pp. 61–78, 2025.
https://doi.org/10.1007/978-3-031-86240-3_5

also present the ongoing work to make ARC available via the European Grid Infrastructure (EGI) Infrastructure Manager.

Keywords: Nordugrid ARC · Galaxy Project · EuroScienceGateway Project · EGI · Distributed computing · Middleware · Grid · Storage and compute · HPC · Cloud

1   Introduction

In this paper we present how the Nordugrid ARC middleware [5,16] has been integrated into the Galaxy web portal [3], and in the EGI [6] Infrastructure Manager [4], both in the context of the EuroScienceGateway project [7]. These integrations have the potential to allow new scientific user-communities benefit from the ARC technology, as the High Energy Physics community have for more than 20 years. With the Galaxy portal integration, job submission to ARC-enabled compute resources can be completely abstracted away from the end-user: a computational job can be sent to an ARC-enabled remote computation site without the user having to relate to ARC or the HPC cluster at all. With the EGI IM integration, a system administrator can easily deploy an ARC-enabled cloud cluster with a ready-made image including a local resource manager system (Slurm or HT-Condor) and an ARC edge service.

In the following we present the Nordugrid ARC middleware, focusing on the main capabilities allowing distributed computing and storage. We furthermore present how the current integration in Galaxy is done, and what are the next steps for a fully production-ready product. We also present the ongoing integration to the EGI IM platform.

2   Nordugrid ARC and the ARC-CE

The Nordugrid ARC middleware was developed as one of several technologies to provide distributed compute and storage capabilities required by the experiments at the Large Hadron Collider (LHC). ARC is installed as an edge service to a computational resource, being a traditional HPC, a dedicated cluster, or a cloud-provisioned resource. The ARC-Compute-Element (ARC-CE) acts as a bridge between the user-requests and the local batch system. To the end-user all complexity of handling different types of compute cluster setups, like OS type, batch system type, software availability, or worrying about moving data in and out of the cluster is abstracted away and handled by ARC. ARC was developed especially for shared HPC systems, where the data is not assumed to be local, and where there might not be outbound connectivity on the compute node. Therefore ARC takes care of all data staging on the edge server: input data is downloaded before the job is handed off to the local batch system, output data is staged out once the job is done and has returned back to the ARC server. In this way valuable compute time is not wasted on data transfers, no additional external software needs to be installed, and no outbound connectivity is needed on the compute nodes. Moreover: ARC caches its input data, avoiding repeated downloads, thus saving data centre bandwidth.

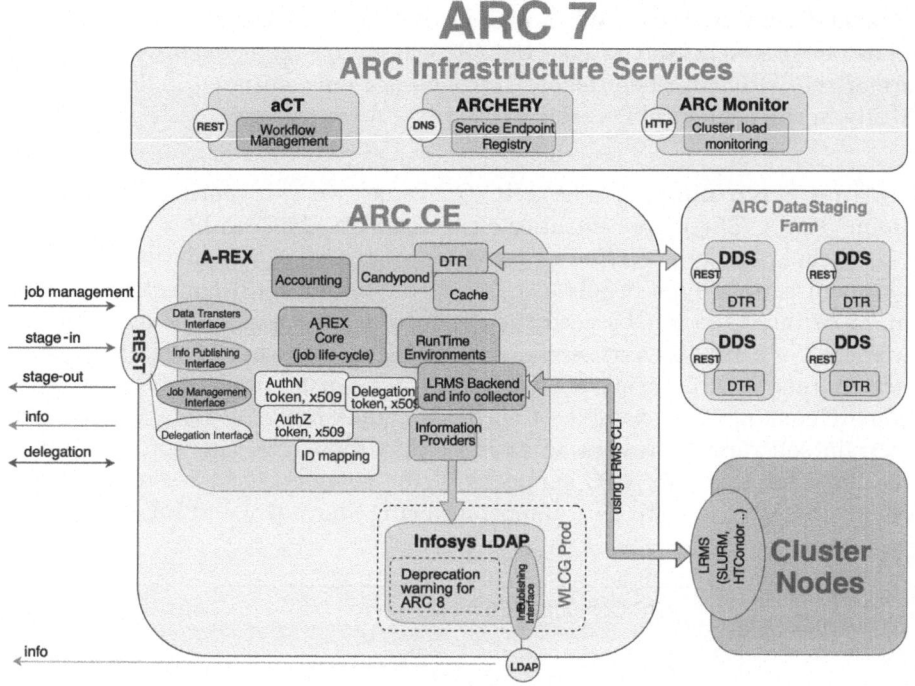

Fig. 1. The ARC-CE components and infrastructure ecosystem.

ARC Components. ARC has undergone many changes since it was first deployed in 2002. Up until recently, ARC supported several different interface types, among them EMI-ES, gridftp and REST. In the current version of ARC we have chosen to focus on the REST interface only, as the others have become less relevant. Also, as the development and support for OIDC token authentication is now well established and expected by users, ARC now supports token authentication. In addition we keep supporting the traditional X.509 [12] authentication which is still used by the high energy physics communities. Figure 1 shows the current ARC components in addition to 3 optional ARC Infrastructure Services. As we see, the REST interface handles the requests for job management (submission, cancellation, retrieval etc.), stage-in and -out of data, resource and job information request in addition to authentication delegation for storage services.

A-REX and Job Lifecycle. In the ARC-CE it is A-REX (ARC Resource-coupled EXecution service) that takes care of the job as it goes through its life cycle from initial request from the user, through handling input-data, submitting the job to the underlying batch system, and marking the job as done once finished and ready to be fetched by the user. A-REX also ensures that any output data that is to be staged out remotely, will be handled by the Data Delivery Service.

Data Delivery Service. When it comes to the data delivery service, this can run as one or several separate servers, depending on the need, or can be run as part of the ARC-CE. The remote transfers are only initiated if the file is not already in the configurable cache.

Runtime Environment. The ARC-CE comes with a Runtime Environment system (RTE). These are central or custom scripts that can be run on demand or by default for jobs. RTE's are used for customizing job runtime needs, for instance if certain type of jobs need a set of specific environment variables to run, or certain type of jobs need specific LRMS options not already defined by the user. They are also very useful in connection with the Information system, since runtime environments are published and can be used by an ARC client for matchmaking. An example of such matchmaking could be a job requiring a certain software: only sites advertising that they have this software will be considered for this job. ARC supports all the common local resource manager systems (LRMS), the two most important being Slurm [15] and HT-Condor [11].

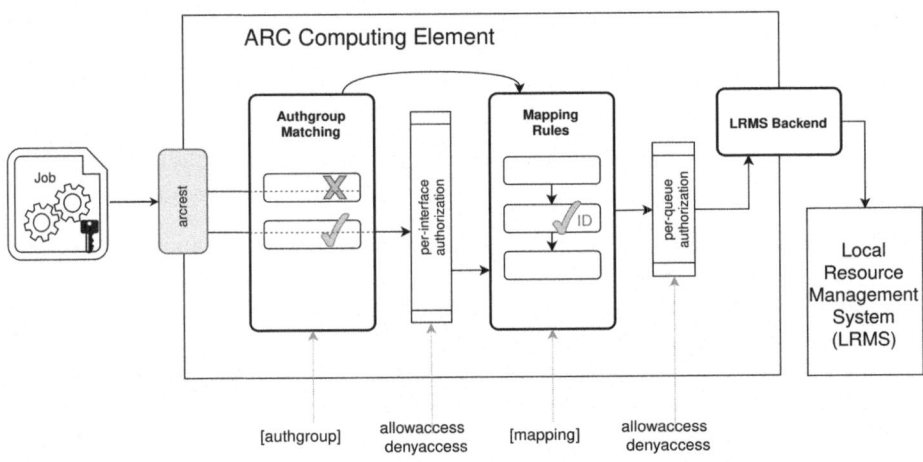

Fig. 2. The configurable ARC authentication, authorisation and user-mapping.

Access Control. The system administrator configuring ARC has full and flexible control over what groups or individuals access to the ARC services through the authentication group configuration settings. These are based on OIDC token or X.509 certificate properties. The authentication and authorization functionality is shown in Fig. 2 together with the inbuilt user-mapping. Both authorisation of job management and to file access is controlled by the authentication groups (authgroups) configured in ARC and matched with the users OIDC token or proxy certificate. For example, the ARC admin can configure an authgroup which allows all jobs that are submitted with tokens issued from the WLCG IAM. Or only tokens with certain scopes or claims from certain token issuers.

Multiple authgroups can be defined. If a users token or x509 user certificate does not match any authgroups, the users job will be rejected. The authentication groups are in turn used to control access to the services, in addition to controlling the user-mapping. Different authgroups can be mapped to different users, and a plugin-functionality allows scripts to handle more complex user-mapping. Access to services and to batch system queues can be controlled by means of the authgroups.

The users token or proxy-certificate follows the job through its life cycle, ensuring full trackeability. This is important on shared resources such as HPC systems, both for security reasons, but also for accounting purposes. The job itself is submitted as the mapped user, fully controlled by the system administrator with the settings in the ARC configuration file.

ARCHERY Infrastructure Service. Of the infrastructure services of relevance to this work is ARCHERY [13], which is a DNS based tool for service discovery and ARC service registry. An ARCHERY registry can be per country or all the way down to per project or community, and holds an indexed list of service endpoint information in that domain. This registry is then used by the ARC client to select an appropriate ARC-CE for the job. A community running an ARCHERY registry can use this as an "all-in-one" registry that holds both a list of ARC-CE resource endpoints, and execution software environments in form of RTE's that the community uses. In the context of Galaxy, one could set up such an ARCHERY registry with all the ARC-enabled resources that the Galaxy users have access to, which then can be used for matchmaking and brokering purposes. This is discussed further in Sect. 5.

3   The Galaxy Web Portal

The Galaxy Project [3, 10] web-portal is a powerful open-source platform used for data-intensive scientific research. It provides an online environment for scientists to perform computational and data analysis tasks efficiently and reproducibly. The key goal of the Galaxy Project is to simplify and streamline the data analysis process by making complex tools and workflows accessible to researchers without the need of any programming skills. Users can import their own data or access publicly available datasets, and then perform a series of manipulations, analyses, and visualizations on that data, all from a simple web interface. The bulk of the tools are in the field of bioinformatics, as Galaxy was created specifically for the genomics community, however, as its core functionality is domain agnostic, other communities have embraced it and have created their own flavour with domain specific tools [8, 9, 17, 18].

By providing a web-based interface, the Galaxy Project eliminates the need for researchers to install and configure complex software packages on their own computers, making it much easier to access cutting-edge analysis tools. Moreover, Galaxy ensures research reproducibility by saving complete histories of each analysis, including the tools used and their parameter settings, allowing researchers to easily reproduce and share their analysis workflows.

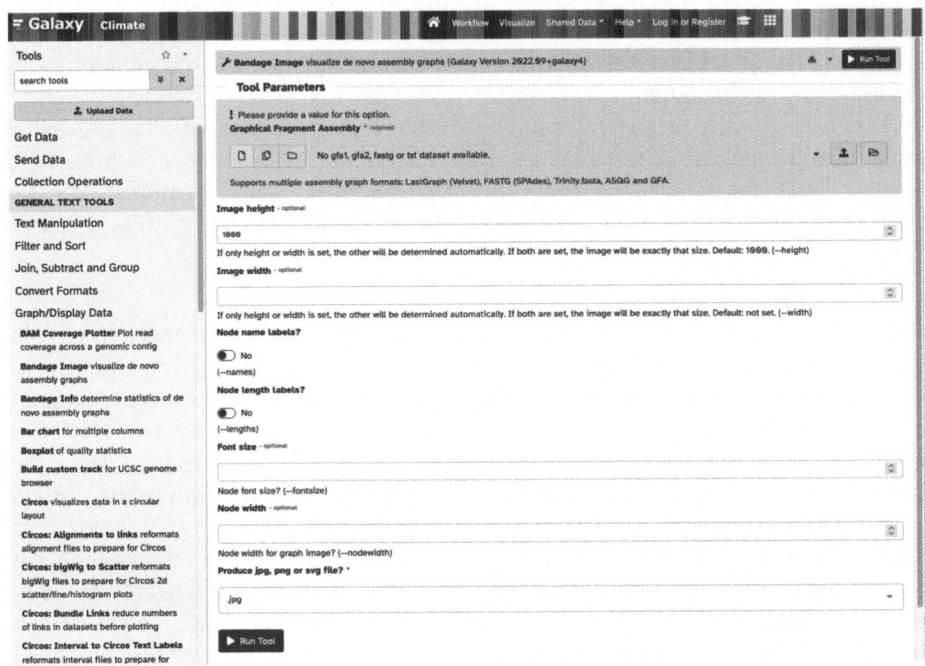

Fig. 3. Screenshot of the climate/usergalaxy.eu web portal, showing a typical tool input parameter page. On the left you see the list of tools available for this portal, in addition to the button to upload data into the Galaxy portal which in turn is used as input-data for the tools.

Galaxy servers are organized either by country or geographical area, like the usegalaxy.eu or usegalaxy.no or domain-specific like https://climate.usegalaxy.eu/ and https://cheminformatics.usegalaxy.eu/ to mention just a couple. There are also many smaller local Galaxy servers being run for instance at universities. Each server will typically have a local compute cluster connected to it, or one or several cloud clusters. When a user submits a job, Galaxy takes care of translating the job input parameters to the submission language used by the cluster, which can be for instance an HT-Condor local resource management system, or a Kubernetes cluster. It is the so-called job-runners described in Sect. 4 that do this job.

An example of how a Galaxy web-interface may look like is shown in Fig. 3, this is the climate/usegalaxy.eu server. Here a tool has been opened showing how the user interacts with the different input-fields needed for that tool. On the top left you can see a button for uploading data into the Galaxy server, that then can be linked to the job.

Previously, users with computational time on HPC systems have not been able to submit jobs to the HPC through Galaxy. This has mainly been because proper authentication required by the HPC could not be provided by Galaxy. Now that ARC is integrated with Galaxy, using the inbuilt ARC authentication and authorization as described above, submission to HPC infrastructures is possible. The integration is described in the next section.

4   ARC Integration with Galaxy for the EuroScienceGateway Project

The integration of ARC with Galaxy is done specifically by creating and enabling a so-called ARC job runner. Galaxy job runners are pluggable, configurable components that take care of the communication with the different types of compute resources that Galaxy supports. The runners can be seen as Galaxy client wrappers for the different flavours of compute resources and provide mechanisms to communicate with the underlying cluster. Examples are the traditional local batch system runners such as Slurm or HT-Condor, a dedicated runner for Kubernetes clusters, one for a so-called PULSAR remote resource, and now the new ARC runner, see Fig. 4.

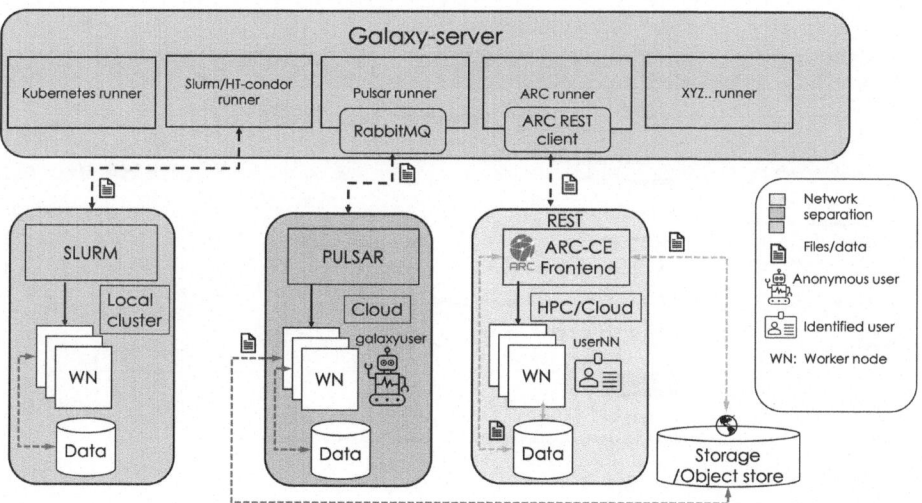

Fig. 4. Galaxy with a few runners showing the integration to different infrastructures. The yellow box shows the ARC integration. (Color figure online)

Galaxy Remote Execution. There are two runners that allow *remote* job execution, where remote is meant in relation to the Galaxy server: Pulsar and the new ARC runner. Pulsar runs on the remote site, typically a cloud resource, and *pulls* jobs from a dedicated and secure message queue between Galaxy and

Pulsar. The input-data is fetched from Galaxy by Pulsar, or from a remote object storage as part of the job when it is executed on the compute node. All jobs run as a single pre-defined user which is not related to the original Galaxy user submitting the job. This model is well suited for dedicated cloud clusters.

ARC Push Model. While Pulsar pulls jobs onto the site, the ARC job runner *pushes* the job onto the remote site, selecting a site according to the user preferences in Galaxy (for instance the users favourite HPC where he has an account), or according to matchmaking rules among sites published in for instance an ARCHERY service or the Galaxy native metascheduling system[1]. Input-data is fetched on the edge ARC service and not on the compute-nodes, and the job is run as the actual user submitting the job. This way of running the job is especially suited for shared HPC sites, as explained.

ARC Job-Runner Job Submission Test. As mentioned, the job runner is the interface between Galaxy and the submission client, in this case the ARC client. The job runner composes a job description based on the Galaxy tool input parameters selected and filled in by the user through a user-friendly web interface as explained in Sect. 3 Each tool in Galaxy is an interface to a software that will ultimately be executed on the underlying compute resource. An example of such a tool, is the very basic test-tool for ARC shown in Fig. 5.

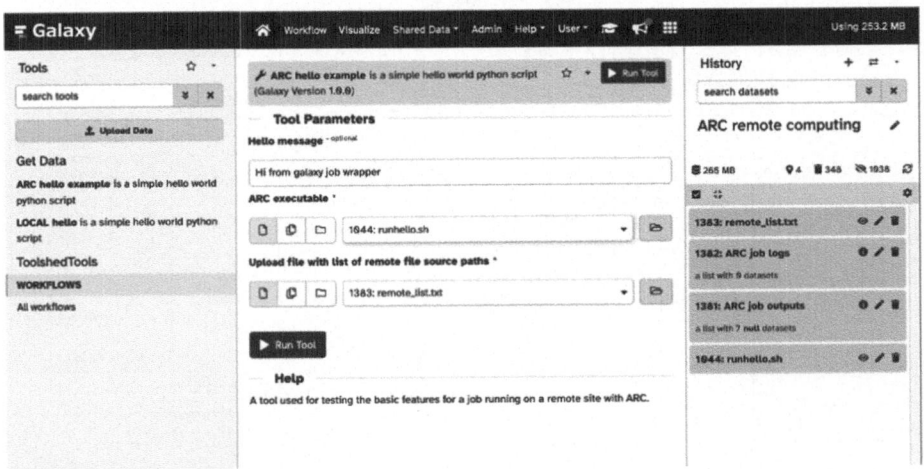

Fig. 5. Demo tool to show how to send a "hello world" job to ARC via Galaxy.

This test-tool takes 3 input parameters: A message in text-form, a script to run, in addition to a text file containing a list of urls to remote input files.

---

[1] In this scenario one would most probably have some dedicated galaxy partitions on the HPC systems where all users from Galaxy were allowed to submit to avoid job submission rejections which would be confusing to the user.

The job runner will translate these input parameters to the ARC job description language in xRSL [1], and then submit the job to an ARC compute resource using the ARC REST interface. Note that tools in Galaxy are not runner-specific, any tool can be run with any runner. The purpose of the example tool shown here is to have a stripped down tool that can showcase the basic functionalities of ARC. However, all tools that Galaxy offer will in principle work (as they do on all the other runners).[2] The test-tool is just a demonstrator to show that a set of input-parameters are passed on and translated to the ARC job description language, the job is then sent to the remote ARC-CE, ARC then downloads the remote input data, and finally the executable is run on the local batch system of that remote resource, taking the input-message as a parameter to the executable. The output of the job is fetched by Galaxy once it is done, and appears in the Galaxy history for the user to inspect, see Fig. 6.

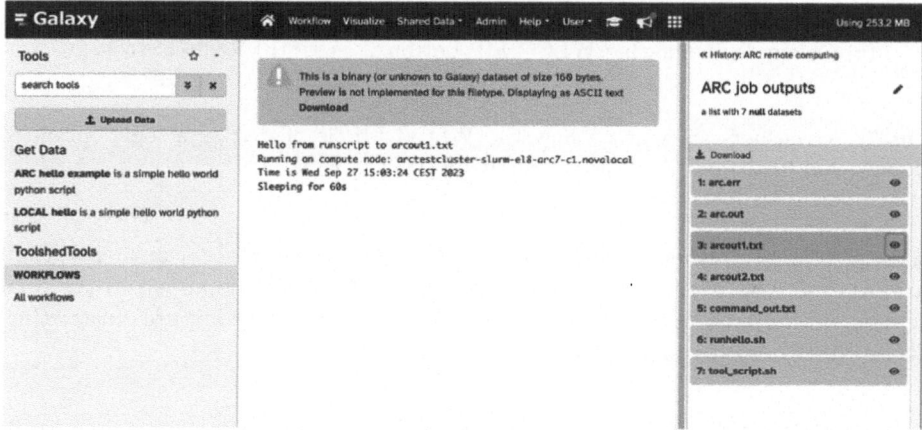

Fig. 6. Output returned by the demo tool once a job has been submitted to ARC.

Authentication. The remote ARC sites require user authentication either via X.509 certificates or via tokens from supported identity providers (IdPs). One such supported IdP is the WLCG IAM [19]. A backend for the WLCG IAM was integrated into the social-core package [14] used by Galaxy for the OIDC authentication. A Galaxy instance can then enable this authentication method (and create an OIDC client for WLCG IAM), which allows the user to log into Galaxy using credentials connected to WLCG (Fig. 7). The ARC job runner then

---

[2] Also note that the example tool is different that standard Galaxy tool as one would not upload an executable to run, rather each tool defines what software to run, and it will either be present on the compute resource or be downloaded from an external repository by the job itself, for instance as an image. Also, in the final version of the job runner, the remote input-files needed will specified in a more integrated way in Galaxy, and not as a simple text-file uploaded with the job.

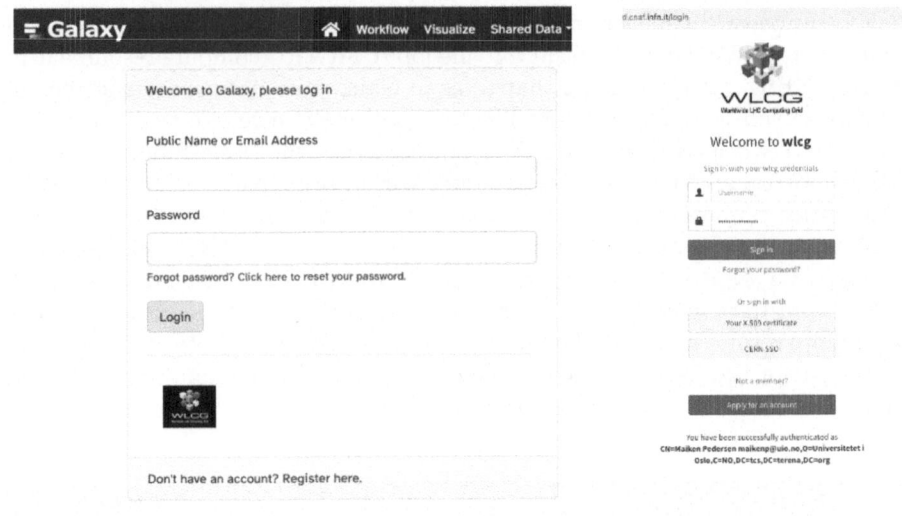

(a) Galaxy login showing WLCG IdP button   (b) Redirect to WLCG
                                                      IdP

Fig. 7. Screenshots showing a) how the WLCG IdP is offered as a login-option to Galaxy, and b) the authentication step at the WLCG IAM.

uses the token refresher mechanism to send a fresh token used for authentication towards the ARC remote computing site. Other IdPs will be added in the final version, allowing users to select the appropriate IdP for the compute cluster they would like to use through Galaxy.

Selecting Remote ARC Site. Finally, in this first prototype version, the user manually provides what ARC-CE remote resource to connect to by setting a user preference. No other configuration is needed from the user point of view, as it is the users token that decides if the job is accepted or rejected on the ARC-CE as explained in the *Access control* paragraph in Sect. 2.

5   Future Work on the Galaxy ARC Job Runner in the Context of the EuroScience Gateway Project

The first bare minimum prototype of the ARC job runner is already in place as shown in Sect. 4. For a fully production-ready version there are several enhancements and additions that need to be made, which will be discussed below.

Remote Paths. As mentioned, ARC handles all the input and output data itself, and thus does not rely on any data handling by Galaxy. The datastaging happens on the remote ARC site, and all file paths used when running the job are relative to the remote site, not to the Galaxy site. For this to work properly, a remote path mechanism is needed so that the correct paths are ingested into the executable script produced by Galaxy[3]. A similar mechanism is used by the remote Pulsar runner. In the current prototype a simple path-rewrite is done with limited functionality in order to get the demonstrator tool working. With the full remote path-handling, ARC will be able to consume the whole set of tools available for Galaxy.

Output Data. Galaxy allows complex rules and filtering related to output data of the job. In existing Galaxy job-runners this is performed as part of the job on the compute cluster, and this is not practical for the ARC client which is responsible for fetching the jobs output files back to Galaxy. The reason is that the ARC client already at submission time expects to know either the detailed list of output files to fetch, or fetch all output files from the working directory of the job. For this to work well with Galaxy, the default behaviour in the final version will be that all the output files will be archived and fetched back into Galaxy, and once in place, the rules and filters will be applied on the unpacked archive and appropriately displayed to the user in the Galaxy web interface.

Dynamic Authentication. Another important enhancement needed in Galaxy is a more flexible way of authenticating towards the selected remote ARC resource. A typical ARC-enabled compute infrastructure has a well defined set of authentication and mapping rules in place, as described in Sect. 2. This means that for instance only tokens issued by certain IdPs are accepted, and different ARC sites will have a different set of accepted IdPs. Depending on the selected ARC remote resource, the user will need to dynamically authenticate before the job is sent, which again means that Galaxy needs to be able to manage several access tokens at the same time.

Brokering and Matchmaking. A related point is the dynamic selection of ARC endpoints based on matchmaking between the tool requirements, input data placement and size, in addition to actual access to the remote ARC site for this particular user. ARC publishes information about the site by the inbuilt information system [2] as we also saw in Fig. 1. Examples of types of information published is the cluster architecture, OS, batch system type, max number of jobs the site can take, current number of running and queueing jobs according to ARC, what Runtime Environments the site has enabled, what Certificate Authorities that are installed, and much more. During submission time an eligible site among the sites the ARC client is brokering, for instance based on available software published through the Runtime Environment is selected. Input data locality can also be taken into account by features currently being developed in a Galaxy metascheduling layer. When it comes to the ARC job runner seamlessly selecting an ARC site among the sites the user has submission access to, two

---

[3] This executable script contains the commands to run the software of the job workload, the paths to the input data and so on.

use-cases cover the most probably scenarios. The first is when there is only a couple of sites: the user would as now enter the sites in his user preferences, and from these sites the most appropriate would be selected by the ARC job runner as described above. If there are many sites, the most practical solution would be to register these sites in a community registry with ARCHERY, and let the registry functionality in the ARC client expose these sites. The user would be a member of this community, and his x509 certificate or token would carry this information, which in turn will be used by the ARC server to authenticate the user. A combination of the ARC and Galaxy brokering and matchmaking mechanisms should be in place in the final version of the ARC job runner.

Once these components are all in place, the Galaxy portal will be a fully fledged distributed computing job submission portal.

6   Integration in EGI Infrastructure Manager

Another initiative that will contribute to a wider set of user-communities becoming aware of the benefits of ARC, is the ongoing integration with the EGI Infrastructure Manager [4] (EGI IM). EGI was a development of the European Data-Grid project founded in 2001, mainly motivated by the storage and compute needs required by the LHC. Today EGI has evolved into a large set of services for federated compute and storage serving "research and innovation", and reaches beyond the LHC.

The EGI *IM* is a service advertised as an open-source solution for creating "complex and tailored virtual infrastructures across multiple clouds". The ARC integration in EGI IM allows users to provision an ARC-enabled Slurm or HT-Condor cluster through the IM web interface. This provides a very straightforward way of setting up an ARC cluster requiring just a few clicks and some minor configuration settings via the web interface.

6.1   Deploying an ARC Enabled Slurm Cluster in EGI IM

In the following we will show the steps necessary to set up a Slurm HPC cluster with an ARC edge service. The implementation is a prototype, and for now only Slurm is available, however, in the future more of the batch systems that ARC supports can be included. 4 main steps are needed to have a fully ready ARC enabled Slurm HPC cluster. These steps assume that the user has created an EGI Check-in account[4], and is a member of a Virtual Organization[5] with enough cloud resources to deploy ARC.

First, the user accesses the Infrastructure Manager at https://im.egi.eu/ using an EGI Check-in account, and from there configures the credentials to be used in Infrastructure Manager[6] (Fig. 8).

---

[4] https://docs.egi.eu/users/aai/check-in/signup/.

[5] https://docs.egi.eu/users/aai/check-in/joining-virtual-organisation/.

[6] https://docs.egi.eu/users/compute/orchestration/im/dashboard/#cloud-credentials.

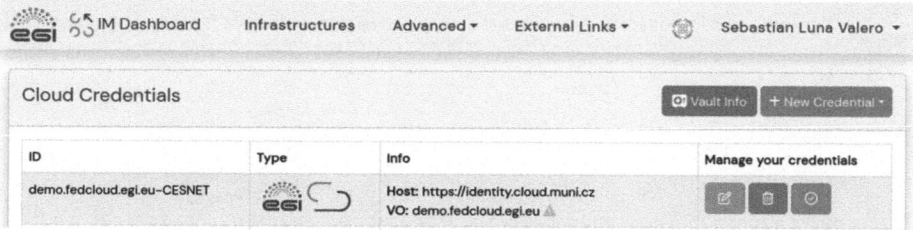

Fig. 8. Configure credentials in Infrastructure Manager.

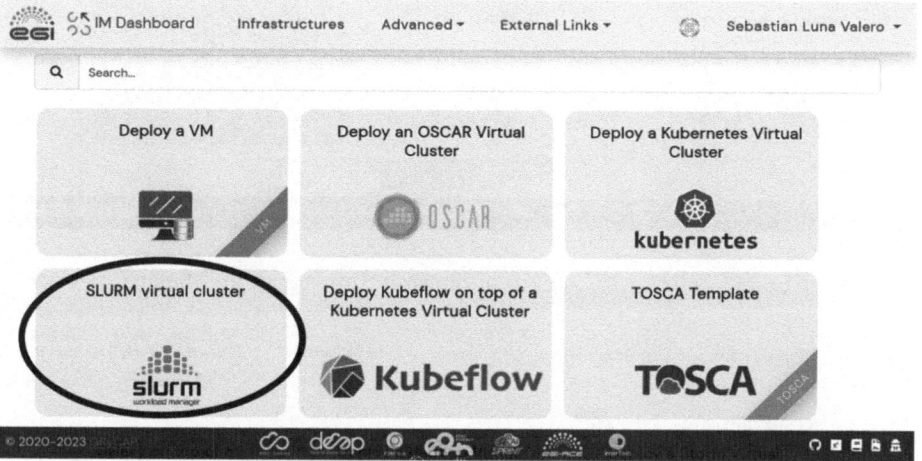

Fig. 9. Select Slurm in Infrastructure Manager.

After this, the user is ready to start deployment. For the ARC Slurm cluster setup, the user selects Slurm from the IM dashboard https://im.egi.eu/im-dashboard/ as shown in Fig. 9. In the next window ARC is added to the infrastructure (Fig. 10). The next step is configuration of the cluster as shown in Fig. 11, both the frontend-end and worker-node features (Number of CPUs, RAM, Cloud Provider and Virtual Machine Image), in addition to a handful of configurable Slurm and ARC features (Slurm feature, ARC timezone).

Finally the cluster is deployed and ready to test (Fig. 12). We see that the IM shows the: 1) the name of the deployment, 2) a unique ID for it, 3) the cloud where the deployment happened (with additional information about the specific site and VO, in the case of the EGI Federated Cloud), 4) status of the deployment, 5) information about the number of Virtual Machines deployed, and 6) the Actions menu.

To log into the cluster with ssh and test the setup, the *Outputs* button which can be seen in Fig. 12 provides necessary information: the username, the public SSH key, and the public IP address.

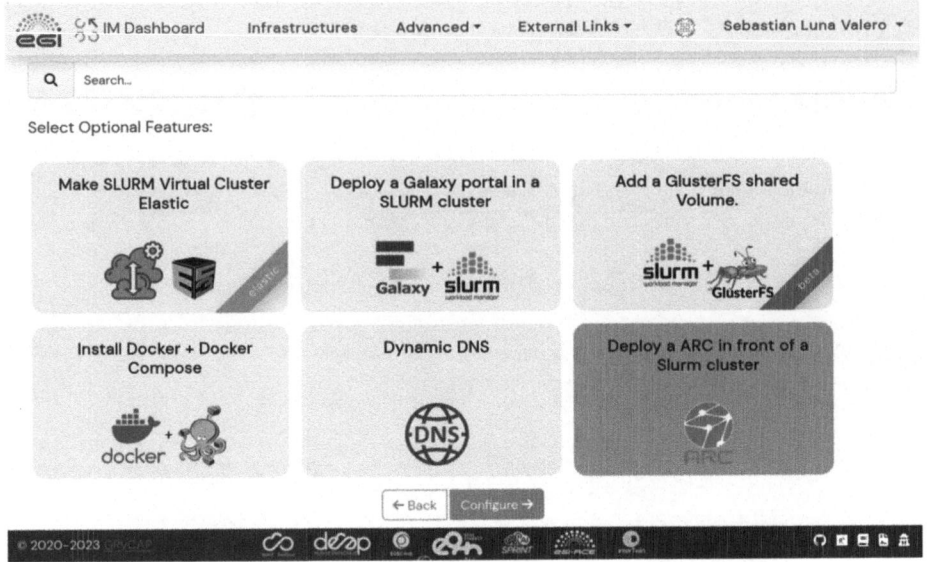

Fig. 10. Select ARC on top of Slurm in Infrastructure Manager.

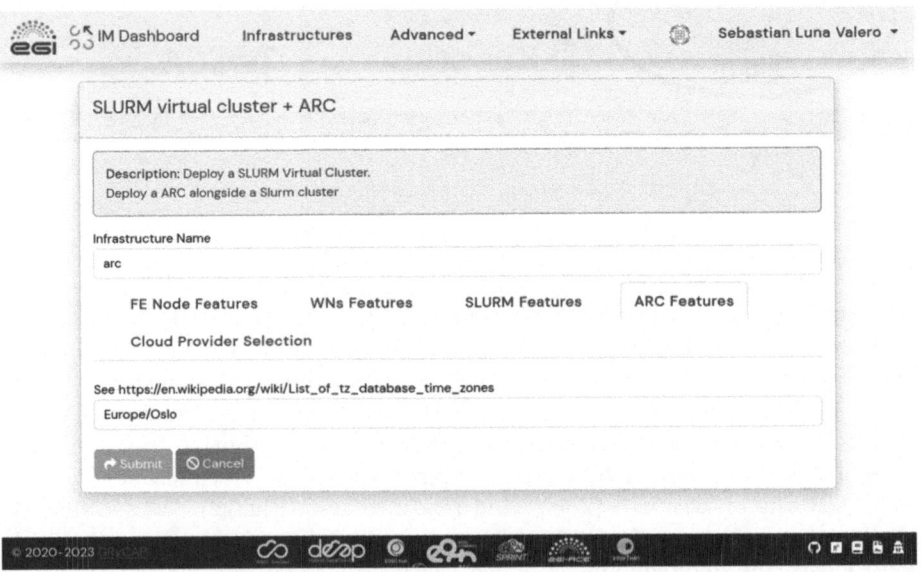

Fig. 11. Configure the deployment with Infrastructure Manager.

6.2   Testing Job Submission

To test the ARC and Slurm deployment on EGI IM, we submitted a hello-world test-job from a remote standalone client to the ARC EGI server using the ARC

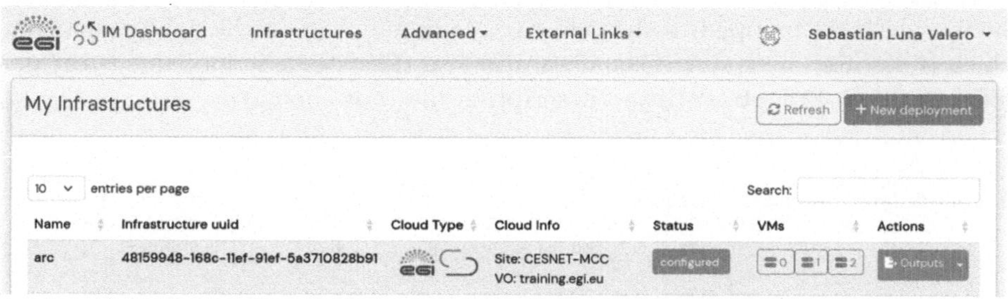

Fig. 12. Check the status of the deployment in Infrastructure Manager.

client command **arcsub** as shown in Fig. 13. Before doing so, we created a DNS record for the server and named it **arc-egi-test.cern-test.uiocloud.no** to avoid having to submit using the ip-address. We authenticated ourselves by submitting the job with a token we know that the ARC-CE accepts[7]. We see that the job was submitted successfully at least from the ARC client point of view. To verify that the job indeed did run as expected on the EGI ARC server, we log in with ssh and check Slurm and ARC.

```
[almalinux@arc-client-neic24-demo testing]$ arcsub -C arc-egi-test.cern-test.
uiocloud.no hellojob.xrls


Job submitted with jobid: https://arc-egi-test.cern-test.uiocloud.no:443/arex
/rest/1.0/jobs/3344b58e23c6
```

Fig. 13. The job is submitted from a remote ARC client server to the EGI ARC server.

On the ARC-CE **arc-egi-test.cern-test.uiocloud.no** we could verify that the job indeed was submitted and running according to ARC. Figure 14 shows the ARC job information.

```
[cloudadm@arc-egi-test ~]$ sudo arcctl job info 3344b58e23c6
Name            : hellojob
Owner           : https://arc-client-neic24-demo.neic24.uiocloud.no/arc/testjw
t/bd4d848e/1000/Test User 13653818
State           : INLRMS
LRMS ID         : 10
Modified        : 2024-05-23 14:15:03
```

Fig. 14. The ARC job information on the EGI ARC-CE.

Furthermore, we see that the job is running on the EGI workernode **vnode-1** shown from the Slurm batch system **squeue** command in Fig. 15. We see that

---

[7] In this case a token issued by a test-token issuer trusted by the ARC-CE.

the job has been mapped to the user `cloudadm` and runs in the `debug` partition, which is dictated by the ARC configuration on this site. The jobname is the one given in the `hellojob.xrls` job description file (not shown).

```
[cloudadm@arc-egi-test ~]$ squeue
            JOBID PARTITION      NAME      USER ST       TIME  NODES NODELIST(REASON)
               10     debug hellojob cloudadm  R       0:04      1 vnode-1
```

Fig. 15. Showing that the ARC job is running on the EGI ARC Slurm cluster as the cloudadm user in this case as per arc configuration settings.

With this, we have verified that IM provides a fully functional minimal ARC setup ready for test-job submission. For a production ready system, changes to the configuration are needed, and that would be done as for a normal ARC-CE: by logging into the machine via ssh, and manually changing the configuration file. One of the important customisation an administrator of the ARC-CE will always want to do for a production-ready system is to set up the authentication groups and their access to the ARC services as described in Sect. 2.

Again, as mentioned, ARCHERY can be set up to hold a registry of the different ARC-CE's related to countries/communities/groups, and in that way the ARC-CE can be integrated into a distributed computational grid. With the ARC job runner in Galaxy, these cloud clusters can of course also be used as computational end-points for jobs submitted through the Galaxy portal.

7   Conclusion

In this paper we have described the ongoing technical developments in Galaxy and in Nordugrid ARC, in the context of the EuroScienceGateway project. The result is the possibility to provide distributed computing to new user groups other than the LHC High Energy Physics community, namely the communities that today or in the future use a Galaxy portal, and have access to HPC infrastructure. They will with this development be able to do all their work from Galaxy, even when wanting to run analysis jobs requiring heavy input data on an HPC, provided the HPC center is ARC-enabled. We have described how the integration of ARC with the Galaxy portal allows Galaxy users to submit jobs to their ARC-enabled HPC systems through the portal. We have also described how integration into the EGI IM allows ARC-enabled cloud clusters to be easily provisioned, and as with the ARC-enabled HPC clusters, can also be used for computational jobs submitted from the Galaxy portal. Both the ARC runner prototype and the EGI IM prototype have been tested and work well.

The development is still ongoing, and will continue to during the period of the ESG project ending August 2025. In the end the goal is to have full and production-ready integration of ARC in Galaxy and in EGI IM.

References

1. Globus Alliance. Extended resource specification language (xrsl). Technical Report (2009)
2. ARC Information System Technical Details. http://www.nordugrid.org/arc/arc7/tech/infosys/infosys.html (visited on 05/23/2024)
3. The Galaxy Community. "The Galaxy platform for accessible, reproducible and collaborative biomedical analyses: 2022 update". In: *Nucleic Acids Research* 50.W1 (Apr. 2022), W345–W351. issn: 0305-1048. https://doi.org/10.1093/nar/gkac247. eprint: https://academic.oup.com/nar/article-pdf/50/W1/W345/45189566/gkac247.pdf
4. EGI Infrastructure Manager. https://www.egi.eu/service/infrastructure-manager/ (visited on 03/25/2024)
5. Ellert, M., et al.: "Advanced Resource Connector middleware for lightweight computational Grids". In: Future Generation Computer Systems 23.2 (2007), pp. 219–240. issn: 0167-739X. https://doi.org/10.1016/j.future.2006.05.008, https://www.sciencedirect.com/science/article/pii/S0167739X06001178
6. European Grid Infrastructure. https://www.egi.eu/ (visited on 03/25/2024)
7. Eurosciencegateway project. https://galaxyproject.org/projects/esg/
8. Galaxy for Astrophysics. https://galaxy.odahub.fr/ (visited on 05/23/2024)
9. Galaxy for Materials Science. https://materials.usegalaxy.eu/ (visited on 05/23/2024)
10. Galaxy Project. https://galaxyproject.org/ (visited on 03/25/2024)
11. Ht-condor. https://htcondor.readthedocs.io/en/latest/
12. ITU. X.509, 02 2011. http://www.itu.int/rec/T-REC-X.509
13. Nordugrid ARC ARCHERY. https://www.nordugrid.org/arc/arc7/admins/archery/index.html (visited on 03/25/2024)
14. Python Social core package. https://github.com/python-social-auth/social-core (visited on 03/25/2024)
15. Slurm. https://slurm.schedmd.com/
16. The Nordugrid Collaboration: ARC7 documentation. http://www.nordugrid.org/arc/arc7/ (visited on 03/25/2024)
17. Usegalaxy ChemInformatics flavour: https://cheminformatics.usegalaxy.eu/ (visited on 05/23/2024)
18. Usegalaxy Climate flavour: https://climate.usegalaxy.eu/ (visited on 05/23/2024)
19. WLCG IAM. https://wlcg-authz-wg.github.io/wlcg-authz-docs/ (visited on 03/25/2024)

Structured Management of Dynamic Geodata - Implementation Guidelines Based on the FAIR Principles

Morten W. Hansen[1]([✉]), Lara Ferrighi[1], Bruce Hackett[1], Øystein Godøy[1],
Johannes T. Langvatn[1], Nina E. Larsgård[1,2], Trygve Aspenes[1],
Arnulf Heimsbakk[1], Charlie Negri[1], Martin G. Pejcoch[3], Shamly Majeed[1],
Elinah K. Kuya[1], Elodie Fernandez[1], Eivind Støylen[1], Vegar Kristiansen[1],
and Audun Christoffersen[1]

[1] Norwegian Meteorological Institute, 0313 Oslo, Norway
mortenwh@met.no
[2] University of Oslo, Section for Meteorology and Oceanography, Oslo, Norway
[3] Lucidia AS, Oslo, Norway

Abstract. Based on the principle that data are only of societal benefit if they are extensively used, MET Norway aims to organize its data
management to support simple, secure and stable access to dynamic
geodata for everyone - from professional users to the general public.
Improving the ability for machine discoverability, interoperability and
reuse is of particular importance. The FAIR principles are key to achieving this goal. Yet, smart joint use of different systems and solutions is
a prerequisite to making data FAIR. A data product cannot be FAIR-
compliant without well-functioning support systems around the product.
This paper outlines the guidelines and solutions for the implementation
of the data management at MET Norway, and how they build upon the
FAIR principles. The guidelines are organized in three main categories:
(1) preservation of data, (2) documentation of data, and (3) publication and sharing of data. The underlying data management model and
guidelines are presented and explained, and MET Norway's implementation is summarized.

Keywords: FAIR Principles · Data Management · Dynamic Geodata

1 Introduction

The Norwegian Meteorological Institute (MET Norway) has a leading position
within the delivery of weather forecasts to the public, and receives about 130
million daily requests to the free and open "location forecast API" (https://
api.met.no/weatherapi/locationforecast/2.0/documentation). A principal value
of the institute is that data are only of societal benefit if they are extensively
used. In 2021, the data behind Yr (https://developer.yr.no) were designated a
global digital public good by the Digital Public Goods Alliance, led by UNICEF

© The Author(s) 2025
A. Azab and T. Malkiewicz (Eds.): NeIC 2024, CCIS 2398, pp. 79–95, 2025.
https://doi.org/10.1007/978-3-031-86240-3_6

(UN). As such, it is clear that MET Norway, and other institutes responsible for environmental research and monitoring, provide important information for many decision-making processes and activities in society, and the ability to trace and verify the data based information and its origin is prominent.

MET Norway manages a very large amount of data, both from own production and from outside sources. The main collections are observations from meteorological and oceanographic instruments (in-situ and remote sensing), simulations from numerical models, and analyzed or derived data products. The data describe geophysical and chemical processes that are constantly evolving and are seen in a geographic framework; hence they are referred to as dynamic geodata. The dynamic geodata are nonetheless quite diverse. Their geographic distribution varies from single points (e.g., weather stations) to gridded 3-dimensional volumes (e.g., numerical models), with a growing number of types in between (e.g., recordings from drifting buoys, sondes, radars, satellites, citizen weather stations). In the temporal dimension, data can have sampling frequencies of seconds up to years and update frequencies of minutes to days. In addition, there are data for which sampling and update frequencies are irregular. Lately, a fifth dimension has been added to many numerical model results as they are being run in ensembles. Managing this range of data types in a unified manner is a major challenge.

Moreover, data obtained for research purposes have been used to less extent in operational services or for quality assurance of these. This has particularly been a challenge in northern and maritime regions, where there is a gap in the observing system and data availability, e.g., as noticed by [16]. Often, non-standardized spreadsheets and other unstructured data storage methods are used to store the information collected. In order for such data to be used beyond specific research purposes, harmonization of both data and documentation of data is required.

Trough a process facilitated by FORCE11 (https://force11.org/), the "FAIR Guiding Principles for scientific data management and stewardship" [19] formalized principles of standardized data documentation, publication and preservation. FAIR stands for Findable, Accessible, Interoperable and Reusable. The FAIR principles provide advice for documenting and making data available, and the ability of machines to automatically find and use the data is in particular focus. Joint use of mutually supportive systems and solutions around the data products is a prerequisite to achieving FAIR-compliance.

The FAIR principles were synthesized by a broadly composed group of representatives from academia, industry, funding agencies and publishers of scientific publications. Although it can be argued that the FAIR principles primarily support the reuse of data among advanced users, the rise of artificial intelligence and the ability to manipulate reality clearly show that the principles are also relevant to the general public's ability to conduct fact-checking.

MET Norway has a free data policy and a long tradition for managing data (e.g., https://www.met.no/en/free-meteorological-data/Licensing-and-crediting), but the responsibility has been fragmented with no common strategy, coordination, unified procedures or standards for documentation, which is required for the organization to work efficiently. The main focus has been to manage storage of operational data. However, delivery, sharing and reuse of data have more recently gained prominence, with growing international focus on data sharing and utilization. MET Norway has therefore implemented a data management regime taking into account both human and organizational roles, as well as technical requirements, in line with the FAIR principles. A major task has been to implement a metadata-driven data management regime where metadata are actively used to both document data and facilitate data handling at the institute, with an end goal to better serve the users of the data.

The implementation of a FAIR data management regime is also driven by external forcing mechanisms at several levels. At the national level, MET Norway must comply with both indications of expected behavior (e.g., OECD regulations) and legal frameworks. For example, the Norwegian government has over time promoted free and open sharing of public data through directives in, e.g., Geodataloven ([12]; implemented as https://www.geonorge.no/), which is a national implementation of the European INSPIRE directive. INSPIRE (https://knowledge-base.inspire.ec.europa.eu/legislation/implementing-rules_en) defines a federated multinational Spatial Data Infrastructure (SDI) for the European Union, similar to the National Spatial Data Infrastructure (NSDI; https://www.fgdc.gov/initiatives/framework) in the USA and the United Nations Spatial Data Infrastructure (UNSDI; https://www.unsdi.nl/about-us/unsdi/). The goal is to provide standardized access to data and provide the necessary tools to be able to work with the data in a unified manner and, as such, also support the concept of interoperability.

This paper provides an insight into MET Norway's data management model, including recommendations for FAIR data management and guidelines thereof. As a foundation for the upcoming recommendations and guidelines, we define two main human roles: data producers and data consumers. Data producers are those who create or produce the data by gathering and storing them, whereas consumers are those who use, or consume, the data. Note that the term "data owner" is absent from the data management vocabulary at MET Norway. MET Norway produces, manages and publishes data but, due to its licensing policy, has no right or ability to control the data once they have been copied to other locations. As such, data ownership is not relevant in this context, although systems that insist on the provision of a defined data owner will get the credentials of the data producer.

The basic functional requirements and constraints of MET Norway's data management are presented in Sect. 2, data governance principles are presented in Sect. 3, and Sect. 4 presents basic concepts to support FAIR management of dynamic geodata. Following this, the implementation of the data management at MET Norway is described in Sect. 5, and the conclusion is given in Sect. 6.

2 Basic Functional Requirements and Constraints

The data management at MET Norway is organized and structured in line with the Open Archival Information System (OAIS; [7]) standard, based on the assumption that the work shall be shared with external organizations and users. Figure 1 is a slightly modified version of Fig. 4-1 in [17]. The dashed line connections identify two-way communication paths, whereas the arrows identify the data and metadata flow between the data producers and consumers. The "Data Management Plan" component in Fig. 1 is similar to the Preservation Planning Functional Entity of [17], whereas the "Data Management Organization" component reflects the Administration Functional Entity of [17]. Figure 1 will be revisited in relevant parts of the following text.

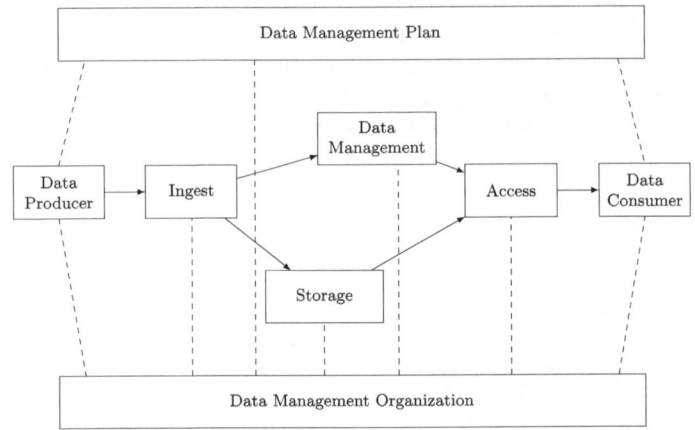

Fig. 1. Data management components at MET Norway, following Fig. 4-1 of [17].

As noticed, the data consumers are at the end of the data management value chain. They may be both external and internal to the institute and, most notably, both humans and machines. The consumers may be users of both data and metadata, and serving them in an efficient, reliable and consistent way, is one of the main motivations for implementing FAIR data management at MET Norway. An important principle is that we do not know the data consumers, and that the data documentation therefore must be extensive and described with rich metadata (FAIR principle F2). The data consumers' needs and requirements serve as the guidelines for determining what data we provide and how.

As well as simplifying the data provisioning workflow, a well structured data management organization in line with Fig. 1 also helps to minimize work efforts and costs. In order to achieve this, some requirements and constraints need to be defined and implemented, but without loosing the flexibility needed to cover the large diversity of assets to be managed. These can be organized in three main categories: (1) Preservation, (2) Documentation, and (3) Publication and Sharing of data.

The category of data preservation includes short and long term management of data to secure access and availability throughout their lifespans. This includes the data life cycle management concept, i.e., the combined planning and execution of the data management (see Fig. 1) from initial storage through to obsolescence and permanent archiving (or deletion) as outlined in Sect. 3. The actual solutions selected for this, depend on expected and actual usage of the data.

The documentation identifies the dataset's who, what, when, where, and how, and shall make it easy for consumers to find and understand data. The documentation category is aligned with FAIR principles F2 and R1, that "data are described with rich metadata" and that "meta(data) are richly described with a plurality of accurate and relevant attributes". The metadata must comply with established international standards, in particular domain-relevant community standards (FAIR principle R1.3). In order to enrich the contextual knowledge about the data, metadata may contain links to other relevant data or metadata. This could be a link to a higher level collection that the data belong to or, where textual description is required, a link to relevant controlled vocabularies and ontologies, in line with FAIR principle I3, that "(meta)data include qualified references to other (meta)data".

To be able to document the lineage and provenance of the data, all actions taken to produce and maintain them must be documented, and the use of the data in downstream systems should be traceable, e.g., by PID/DOI references (in line with FAIR principle R1.2 "(meta)data are associated with detailed provenance").

The relationship between data and information can be maintained through datasets, as described in Sect. 4.1. In order to give the data consumers seamless access to the information content, metadata should follow the data to the greatest extent possible. This involves application of information containers that support standardized storage of both data and metadata, e.g., through utilization of self-describing file formats. If properly done by the data producer, this helps to simplify and automate the publication and preservation of data through services.

MET Norway's documentation guidelines are presented in Sects. 5.3 and 5.5. In particular, an internal metadata model (Sect. 5.5) supports connection to relevant external resources through links, e.g., to the World Meteorological Organization (WMO) Integrated Global Observing System (WIGOS) observation station metadata repositories for space and surface based capabilities (https://space.oscar.wmo.int/spacecapabilities and https://oscar.wmo.int/surface/#/).

The goal of publication and sharing is to make data Findable and Accessible to consumers. Application of standardized approaches and cost effective solutions, e.g., enabling data streaming for use of subsets in data analysis tools, is vital. This can minimize the overhead in dissemination solutions by reducing the transferred data amount, e.g., to let consumers access pixels instead of images or files. Machine interfaces must serve standard documentation and encoding of data following the requirements of the external systems to which data are served (e.g., Geonorge, INSPIRE, WMO), i.e., supporting interoperability (also FAIR principle R1.3). To ensure that that the data consumer receives the same data for a given request at any time of the request, it is mandatory that data and data access are maintained and kept consistent over time. MET Norway's guidelines in Sects. 5.3 and 5.5 ensure that the publication and sharing category is supported.

3   Governance

Data governance is the overall management of the availability, quality, usability, integrity and security of an organization's data. Sound data governance includes a governing body (the "Data Management Organization"; Fig. 1), a defined set of procedures and a plan to execute the procedures. The data governance makes sure that data management guidelines are implemented throughout the organization, that the practices are aligned with the institute's strategic goals, and that the data management regime is subject to timely review, analysis and revision.

An important element in the data governance is the life cycle management of data. This defines how data generated or used in an activity (e.g., a project) will be handled throughout the lifetime of the activity and after the activity has been completed, i.e., what kind of data that will be generated or acquired, how the data will be described, where the data will be stored, whether and how the data can be shared, and how the data will be retired (archived or deleted). The purpose of life cycle management is to safeguard the data, not just during its "active" period but also for future reuse, and to facilitate cost-effective management. The data life cycle management should be planned and documented in Data Management Plan (DMP; Fig. 1) documents, e.g., as outlined in [15]. A Data Management Plan (DMP) is a text document that describes how the data life cycle management will be carried out for data used and produced in specific activities (e.g., research projects). Generally, projects financed by research funding agencies such as the Research Council of Norway (RCN) and the European Commission (EC) require such documentation, but it may also be beneficial in larger internal projects that are part of the entity's core value chain. MET Norway's internal recommendations for DMPs are shortly described in Sect. 5.2.

In order to align and unify the departments and divisions at the institute, MET Norway has organized the data management as a cross-cutting activity, with a governing body pertaining a defined organizational role. This is described in Sect. 5.1.

4   Basic Concepts

In order to implement the FAIR principles in a technical and organizational framework as depicted in Fig. 1, some basic concepts of the data management are important. In this section, the main guidelines to organizing data in datasets are presented (Sect. 4.1), and various types of metadata (4.2), and controlled vocabularies and ontologies are introduced (Sect. 4.3).

4.1   Dataset

A fundamental concept in the data management regime is the dataset:

Definition 1. *A dataset is a combination of data records, such as the numerical values needed to analyze and interpret a natural process, and the related information content.*

Typically, a dataset represents a number of related variables defined in time and space for an intended use. Datasets may be categorized by: (1) Source (e.g., in-situ or remotely sensed observations or numerical model projections and analyses), (2) Processing Level (e.g., raw data, calibrated data, quality-controlled data, derived parameters, or temporally and/or spatially aggregated variables), and (3) Data Type (e.g., point data, sections and profiles, lines and polylines, polygons, gridded data, volume data, or time series).

The dataset can be stored in, for example, a relational database or as flat files. Data having the same characteristics in a given category, and common variable ranges, are normally considered part of a single dataset. Furthermore, the following guidelines are useful to best serve the data through web services: (1) A dataset is defined as a number of spatial and/or temporal variables with common dimensions, (2) a dataset should be defined by the information content and not the production method, and (3) a good dataset does not mix feature types, i.e., trajectories and gridded data should not be present in the same dataset.

This implies that a dataset should be defined based on a "least common multiple" of information content. In this context, consideration of the dataset granularity, which is a constant challenge in distributed data management, is important. The dataset granularity is usually defined by the data producer, who often has a different perspective on the data than a potential data consumer. A relevant example is temperature and salinity measurements from research cruises. These data are often published as one single cruise dataset. However, a typical data consumer may only be interested in the profile of a given variable, in order to aggregate it with the same type of profiles from other cruises in a specific region and time period. By publishing all data from a given research cruise as one dataset, the workload on the data consumer is increased, with little benefit.

An important linked dataset relation is the parent-child relationship, representing a hierarchical relation between a lower-level (child) and a higher level (parent) resource. To maintain visibility of the cruise, parent-child relations between datasets could be established, or datasets could be tagged with a cruise reference. Similar examples could be identified from other domains. Increasing the dataset granularity in this way, is also useful in order to be able to efficiently serve the data through standard web services, e.g., from the Open Geospatial Consortium (OGC). For example, the OGC Web Map Service (WMS) is not able to handle mixed coordinates from, e.g., a model simulation result with variables defined on different vertical coordinates. As such, this data should be separated as linked datasets.

Generally, the highest possible functional granularity (i.e., the most detailed, yet practical, separation of data into datasets) is beneficial for machine action and reuse in, e.g., calibration and validation of numerical models and satellite retrievals. This implies that, e.g., datasets composed of individual measurements should be avoided because it involves impractical management efforts, whereas several stations (at different locations) in the same dataset should be avoided because it requires extra work by a user who is only interested in data from one of the stations. Different use cases and resulting guidelines for deciding the optimal dataset granularity have been discussed in several contexts, such as, e.g., the Research Data Alliance (RDA) Data Granularity Working Group (https://www.rd-alliance.org/rationale/data-granularity-wg/).

Most importantly, however, the consumer needs related to data delivery, security and preservation should be the main guidelines for the dataset definition. This is often case-dependent, and requires iteration and verification over time, and therefore also flexibility in the data management regime to properly handle new dataset versions (see, e.g., [9]).

4.2   Metadata

The concept of metadata is broad and can be somewhat unclear. In practical data management, the discussion often arises about what is needed or not. Especially, data delivery system managers are concerned that the metadata cause a too great volume increase thus slowing down the data delivery. In this context, the concept of linked data, what is stored where and, particularly, clear definitions of the different categories of metadata becomes important. Here, six metadata categories are relevant: (1) Discovery Metadata, (2) Use Metadata, (3) Provenance Metadata, (4) Site Metadata, (5) Configuration Metadata, and (6) System Metadata. Note that there is some overlap between the categories, and that some information may belong to more than one metadata category.

The purpose of **discovery metadata** is to make data Findable. This type of metadata is also called index metadata because of its connection to the library index card. The discovery metadata describe what, where and when, who created the dataset, how to access the data, and potential use constraints. The discovery metadata also link to further information about the data.

Discovery metadata should be embedded in the information containers, i.e., self-describing files, in order for the data consumer to have easy access to the information content. In addition, the discovery metadata must be stored and indexed separately to allow Findability in searchable catalogs.

Examples of discovery metadata standards are, e.g., ISO 19115 [7] and NASA GCMD/DIF (https://www.earthdata.nasa.gov/learn/find-data/idn/gcmd-keywords).

The purpose of **use metadata** is to help the data consumer understand the data, decide if they are useful in a particular context, and reuse them without need for any further communication. As such, the use metadata describe the actual data with standard vocabularies and units, and provide, e.g., the map projection and how missing values are encoded.

To enhance their availability, also the use metadata should be embedded in the information containers. For instance, machine use is not possible without clear and concise information about the data unit, following a given standard.

Examples of use metadata standards are, e.g., the Binary Universal Form for the Representation of meteorological data (BUFR; see [20]) and General Regularly distributed Information in Binary form (GRIB; see [20]) standards used in the meteorological community, and the Climate and Forecast (CF) convention ([5]) developed for gridded data from climate and forecast models and later adopted in other geosciences.

The purpose of **provenance metadata** is to keep track of what has been done with the data by who, i.e., its origin and lineage. This includes information about the entities and people who created the data, the data production activity, any attribution, and version history. Also, the documentation needed to prove that the data are what they appear to be is relevant here, and can be provided as a file check-sum.

Much of the provenance metadata can also be embedded in the storage container but not, e.g., the file check-sum. Some data centers therefore keep provenance metadata in separate storage, using unique and persistent identifiers to link between the information layers.

An example of a provenance metadata standard is, e.g., the PROV Data Model [1].

The purpose of **site metadata** is to inform about the observation context such as, e.g., the observation location, instrumentation, and collection procedures. The site metadata to some extent overlap with discovery metadata but also augment it. The most significant site metadata, such as the geographical location, are typically repeated in the discovery metadata, whereas specific information about the measuring instrument, such as its date of installation, can be kept in the site metadata only. The data consumer should be able to find such information through the discovery metadata, e.g., provided as a site identifier. Site metadata are also used in other contexts, e.g., for designing observation networks. Examples of site metadata are, e.g., the WMO Integrated Global Observing System (WIGOS; [21]) and OGC O&M [13].

The purpose of **configuration metadata** is to provide configuration information to data services. These metadata are used to tune the services for the given data product, e.g., for how to best visualize it in OGC WMS.

The purpose of **system metadata** is to inform about the technical structure of the data management system, and to track changes in it. This includes, e.g., details about data storage, web services and their purpose, interactions between system components, number of users, and who are the technical responsible for the various system components. These metadata are not relevant to the data consumer but are very important to the institution running the data center.

4.3 Controlled Vocabularies and Ontologies

Controlled vocabularies are lists of concepts with specific labels (i.e., entry terms) and definitions, that can often be resolved on the web. They can be used to annotate data records, but also for tagging and/or categorizing assets. Terms can also follow a hierarchical tree-like structure, or provide synonyms and relations to other terms or resources (taxonomies and thesauri), and can be organized in formal ontologies. By providing formal linkages to other vocabularies or resources, concepts can be mapped or translated between different domains and metadata standards.

There are many examples of controlled lists, glossaries, vocabularies (e.g. [5], https://www.earthdata.nasa.gov/learn/find-data/idn/gcmd-keywords), thesauri and ontologies (e.g. [2,11,14]), covering diverse disciplines, contexts and purposes, as well as online services that collect and provide machine-readable resources (e.g., https://vocab.nerc.ac.uk; [8]).

The purpose of providing metadata information using a semantic approach in a scientific data management framework, is to improve the standardization of the data documentation and make it consistent. In this way, discoverability, interoperability, as well as usability of the data is improved. Since semantic resources can be machine readable, they can easily be integrated in tools supporting data creation and metadata provision, as well as in more complex workflows or lookup services.

5   Implementation

The implementation of the data management regime at MET Norway is built upon the principles outlined in Sect. 2. By forcing these requirements and constraints, the work to automatically aggregate data in space and time is simplified and helps to establish automated workflows, e.g., in support of downstream research activities or the emerging European Data Spaces (see, e.g., [4]). In the following sections, MET Norway's implementation with respect to the components depicted in Fig. 1 are presented.

5.1   Human and Organizational Roles

Human and organizational roles mainly constitute the Data Management Organization entity in Fig. 1. As data flow from the data producers to the data consumers, the Data Management Organization ensures handling and delivery of the data in a systematic and cost-effective manner. The Data Management Organization structure is illustrated in Fig. 2a. Its purpose is to coordinate and facilitate Structured Data Management (SDM) undertaken at MET Norway and ensure that the institute is able to fulfill external requirements in a cost-efficient and pragmatic manner. The intention is to offer a unified framework that can be used as a foundation for development of Application Programming Interfaces (APIs), higher level product generation, and various web portals, in line with the FAIR guiding principles, i.e., to help streamline the value chain, rather than interfering with it. In particular, a Data Manager (DM) and a Data Management Group (DMG) facilitate cross-department communication to avoid duplication of work or mutual blocking.

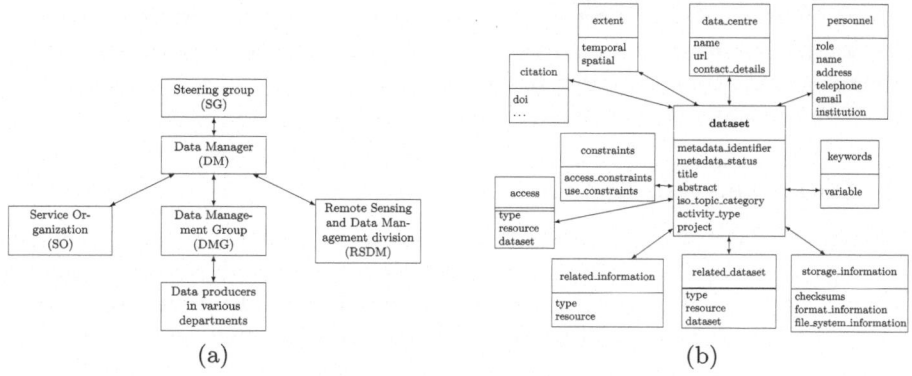

Fig. 2. (a) Data management organization at MET Norway. Coordination is primarily handled by the Data Manager (DM) and the Data Management Group (DMG). (b) Logical outline of the MET Norway MetaData Format [6].

Prioritization, strategic decisions, and resource commitment to the data management work is controlled by a Steering Group (SG) comprising the directors of MET Norway, and headed by the Director General. The Data Manager (DM) has the overall responsibility for ensuring that the institute's data management regime is implemented and followed throughout the organization. The DM coordinates between the different operational and research activities and departments, and ensures that the data management regime is known in the organization.

The Data Management Group (DMG) provides support to the DM, project managers and line managers for the execution of their duties. As such, the DMG members act as local points of contact for data management in the organization. The DMG members contribute relevant competence in IT, numerical prediction (weather, ocean, air quality, climate) and observations (in-situ and remote sensing), and they have hands-on experience with data management in their respective divisions (i.e., data management "super-users"). The DMG has representatives from each relevant department, and its members are nominated by the department heads and approved by the SG. The DM may propose changes to the composition and duties of the DMG as needed, subject to approval by the SG.

The Service Organization (SO) is responsible for operation, technical development and platform management of the services for long-term data. The SO ensures proper implementation and operation of the underlying hardware (processing and storage), as well as the necessary web services and tools, through coordination with relevant actors at the institute, and they develop new functionalities and services in coordination with DMG and the Remote Sensing and Data Management (RSDM) division.

The RSDM division takes care of alignment and coordination with national and international standardization efforts, with participation of other units as required. RSDM performs research and development in line with modern data management, and aligns with relevant developments in the community through participation in standardization activities in national, regional (EU) and global (WMO), as well as discipline-specific systems.

5.2   Data Management Plan

Planning the data management with respect to the data life-cycle is an important aspect of the data governance. Both Norwegian and European funding agencies require submission of a DMP at project initialization, and that an updated version is provided together with the project's final report. The funding agency usually provides requirements and guidelines, or a template, for the DMP. However, if this is not the case, templates based on the recommendations by Science Europe (https://scienceeurope.org/our-priorities/research-data/research-data-management/) can be used. The DMP is prepared by the data producer following guidelines from the data management organization, and user needs. This is important input to, e.g., storage and computing power management.

MET Norway generally recommends a Norwegian tool called EasyDMP (https://easydmp.no/) to customize templates and write DMPs according to the requirements. EasyDMP is not yet machine interoperable in the sense that it can be used automatically in practical management of, e.g., data storage but it contains an export function to download HTML or PDF versions for archiving the DMP. MET Norway is working on the definition of its own template with specific focus on needs for storage and computing power, based on templates from Science Europe.

5.3   Ingest

All the components of MET Norway's data management regime is based on the use of datasets, as defined in Sect. 4.1. Thanks to the self-describing design of the Unidata CDM model ([18]), NetCDF files are recommended as the primary dataset storage containers.

The NetCDF format simplifies the sharing of data externally and is easy to link with standard data sharing protocols like OGC WMS and OPeNDAP. For documentation, the Attribute Convention for Data Discovery (ACDD; https://wiki.esipfed.org/Attribute_Convention_for_Data_Discovery_1-3) and some institute specific ACDD extensions ensure sufficient discovery metadata content, provided as global attributes in the CF-NetCDF files, to ensure interoperability with the most important geographic information schemas (i.e., INSPIRE and the Norwegian profile of INSPIRE, and the WMO Information System, all based on ISO 19115, and NASA GCMD/DIF).

CF compliance (see [5]) ensures adequate use metadata in the CF-NetCDF files. The CF-NetCDF convention specifies a standardised annotation of the dataset variables in the form of CF standard name attributes, compliant with semantic web approaches. Since not all conceivable parameters are covered in CF, the CF governance structure can approve new entry suggestions through discussions in the CF GitHub organization (see https://cfconventions.org/discussion.html).

MET Norway has implemented a number of custom services as part of the Ingest entity in Fig. 1, including basic quality control to support the data producers in publishing their CF-NetCDF datasets. Descriptive information about the datasets (or Archival Information Package; AIP) is extracted and stored in XML files, which are indexed in the Data Management entity and stored (the Storage entity, Fig. 1) along with the CF-NetCDF files. As long as the main guidelines are followed and the standard CF-NetCDF format with the required ACDD metadata are used, the use of these services is straightforward. The structured formatting and documentation of the datasets simplifies the subsequent data management, as well as publication through standard search, access and visualization services.

5.4   Storage

The CF-NetCDF datasets are stored in the Lustre file system (see, e.g., https://www.lustre.org/about/), whereas XML files containing discovery and configuration metadata are stored on local disk systems (e.g., kubernetes persistent volume claims) and version controlled using git to allow semi-automatic maintenance by humans, as well as track control. Both data and metadata storage are maintained by the Service Organization (Fig. 2, left).

5.5   Data Management

The Data Management includes administration and maintenance of the institute specific MET MetaData (MMD; [6]) format, and related query functions and updates. This is handled by the RSDM division. The high level logical structure of the MMD information model is outlined in the Fig. 2(b) with the purpose to illustrate the specific metadata information that are part of the model.

MMD is used for interoperability with GCMD/DIF and various profiles of ISO 19115, which are the most common discovery metadata standards within the geosciences. MMD is also adaptable to new upcoming DCAT-AP standards, and can be extended with provenance and configuration metadata as well as information needed in internal production chains. The compliance requirements imply that the relevant information from GCMD/DIF and ISO 19115 can be transferred into MMD, although not all information in MMD have a place in ISO 19115 or GCMD DIF. As a part of the Storage entity, the git repository with MMD XML files is defined as the authoritative discovery metadata source. These metadata records are then indexed in databases such as PostgreSQL and SolR (i.e., the back-end of the machine readable metadata catalog and the web portal, respectively; see Sect. 5.6), so they can be served in web-searchable metadata catalogs.

Most global (e.g., WMO), regional (e.g., INSPIRE) and national (e.g., Norway Digital) data sharing services rely on ISO 19115 as their discovery metadata structure, although there is a current movement toward DCAT-AP.

A weakness of many top level data sharing efforts is the lack of controlled vocabularies and ontologies for use in free text fields, which can lead to unusable content entries. The community's focus on standardized data documentation and utilization of controlled vocabularies and ontologies is, however, gaining momentum. As such, MMD defines controlled vocabularies that are commonly used at MET Norway, and in the extended meteorological community. This is complemented with, e.g., the CF standard names ([5]) vocabulary, NASA/DIF GCMD keywords (see https://www.earthdata.nasa.gov/learn/find-data/idn/gcmd-keywords), and other relevant controlled vocabularies.

The METNO Vocabulary Server, available at https://vocab.met.no, follows the guidelines of [3] through a dedicated effort in the European ENVRI-FAIR project (https://envri.eu/the-envri-fair-project/). The vocabulary server provides controlled vocabularies defined in MMD for the discovery and use metadata regularly applied by the institute's data producers.

In addition, the vocabulary server provides partial mapping of CF standard names to the NASA GCMD/DIF Science Keywords, that can be used as a look-up service to enrich the metadata with relevant variables and keywords. This is particularly useful for supporting machine interoperability.

Through https://vocab.met.no, MET Norway supports FAIR principle I2 that "(meta)data use vocabularies that follow the FAIR principles".

5.6   Access

The described information structures pave the way for efficient data discovery and use through standard tools and automated services. This includes services for data search and discovery, data access, and data visualization.

Currently, machine interfaces (e.g., https://data.csw.met.no) for data search and discovery are deployed through pyCSW (an OGC Catalogue Service for the Web [CSW] server implementation written in Python; [10]) supporting the Open Archives Initiative - Protocol for Metadata Harvesting (OAI-PMH) and the OGC CSW protocols, and serving metadata following the GCMD/DIF and ISO 19115 standards.

The human interface to the metadata catalog is available via https://data.met.no and is implemented using the Drupal Content Management System (CMS; https://www.drupal.org/) in combination with custom modules linking the back-end and front-end services.

Data access services are available through the OPeNDAP protocol implemented on the Unidata Thematic Real-time Environmental Distributed Data Services (THREDDS) Data Server (TDS). Furthermore, OGC WMS data visualization services provide geo-referenced map images through simple HTTP interfaces.

As long as data producers provide their data according to the guidelines, new services can in principle be added without need for interaction with the data producer.

6 Conclusion

A good data management system helps to minimize efforts and costs, and simplifies the data provisioning work needed to make data available to and reusable for unknown consumers, both humans and machines. In order to achieve this, some requirements and constraints need to be defined and implemented, but without loosing the flexibility needed to cover the large diversity of assets to be managed. As such, MET Norway has established a unified data management regime with the intention of ensuring the availability, ease of use, integrity and security of the organization's data. The data management regime establishes rules and procedures to ensure that guidelines are implemented in the organization, that data management practices are in line with and contribute to the institute's strategic goals, and that the data management regime is reviewed and updated regularly. This is supported and coordinated through the Data Management Organization.

The data management guidelines provide advice to the data producers for how to define and design datasets, how to document data through metadata, and how to store data in order to help simplifying and automating the publication and preservation of data through services, as well as providing seamless data access to consumers. The use of unique identifiers for datasets, required through the ACDD and MMD standards, enables the tracking of information used in decision-making processes, and ensures the ability to verify background data.

Making environmental data open and accessible in line with the FAIR principles is a continuous process. Still, only some of the datasets managed by MET Norway are integrated in the data management regime, and there is a need to follow up the data production systems and people through the Data Management Organization.

Two particular areas that require more work are the documentation of vocabularies used, and establishing the connections between them, as well as the need for a machine interoperable solution for Data Management Plans (DMPs).

References

1. Belhajjame, K., et al.: PROV-DM: The PROV Data Model (2013). https://www.w3.org/TR/2013/REC-prov-dm-20130430/
2. Buttigieg, P.L., Morrison, N., Smith, B., Mungall, C.J., Lewis, S.E.: The ENVO Consortium: The environment ontology: contextualising biological and biomedical entities. J. Biomed. Semantics 4(1), 43 (2013). https://doi.org/10.1186/2041-1480-4-43

3. Cox, S.J.D., Gonzalez-Beltran, A.N., Magagna, B., Marinescu, M.C.: Ten simple rules for making a vocabulary FAIR. PLoS Comput. Biol. **17**(6), 1–15 (2021). https://doi.org/10.1371/journal.pcbi.1009041

4. Curry, E.: Real-time Linked Dataspaces. Springer (2020). https://doi.org/10.1007/978-3-030-29665-0

5. Eaton, B., et al.: NetCDF Climate and Forecast (CF) Metadata Conventions. Tech. rep., CF community (2023). https://cfconventions.org

6. Godøy, Ø., et al.: MET Norway Metadata Format Specification (2020). https://htmlpreview.github.io/?https://github.com/metno/mmd/blob/master/doc/mmd-specification.html

7. Space data and information transfer systems – Open archival information system (OAIS) – Reference model. Standard, International Organization for Standardization, Geneva, CH (2012)

8. Jonquet, C., et al.: Agroportal: A vocabulary and ontology repository for agronomy. Comput. Electron. Agric. **144**, 126–143 (2018). https://doi.org/10.1016/j.compag.2017.10.012, https://www.sciencedirect.com/science/article/pii/S0168169916309541

9. Klump, J., Wyborn, L., Wu, M., Martin, J., Downs, R.R., Asmi, A.: Versioning data is about more than revisions: a conceptual framework and proposed principles. Data Sci. J. **20** (2021). https://doi.org/10.5334/dsj-2021-012

10. Kralidis, T., et al.: geopython/pycsw: 2.6.1 (2023). https://doi.org/10.5281/zenodo.7839119

11. Magagna, B., Peterseil, J., Frenzel, M., Kertész, M., Grandin, U.: EnvThes - Thesaurus for long term ecological research, monitoring and experiments (2013). http://vocabs.lter-europe.net/EnvThes/

12. Norwegian Ministry of Local Government and Regional Development: Lov om infrastruktur for geografisk informasjon (geodataloven) (2010). https://lovdata.no/dokument/NL/lov/2010-09-03-56

13. Open Geospatial Consortium: OGC Earth Observation Metadata profile of Observations & Measurements. Standard, Open Geospatial Consortium (OGC) (2016). https://docs.ogc.org/is/10-157r4/10-157r4.html

14. Publications Office of the European Union: Eurovoc (2024). https://op.europa.eu/en/web/eu-vocabularies/th-dataset/-/resource/dataset/eurovoc

15. Europe, S.: Practical guide to the international alignment of research data management (2021). https://doi.org/10.5281/zenodo.4915861

16. Smith, G.C., et al.: The WWRP PPP Steering Group: Polar ocean observations: A critical gap in the observing system and its effect on environmental predictions from hours to a season. Front. Marine Sci. **6** (2019). https://doi.org/10.3389/fmars.2019.00429, https://www.frontiersin.org/articles/10.3389/fmars.2019.00429

17. The Consultative Committee for Space Data Systems: Reference model for an open archival information system (OAIS) – recommended practice. Standard, CCSDS Secretariat, Space Communications and Navigation Office, 7L70, Space Operations Mission Directorate, NASA Headquarters, Washington, DC 20546-0001, USA (2012). https://public.ccsds.org/pubs/650x0m2s.pdf

18. Unidata: Unidata's common data model version 4 (2014). https://docs.unidata.ucar.edu/netcdf-java/4.6/userguide/CDM/index.html

19. Wilkinson, M.D., et al.: The fair guiding principles for scientific data management and stewardship. Sci. Data **3**, 160018 (2016). https://doi.org/10.1038/sdata.2016.18

20. World Meteorological Organization (WMO): Manual on Codes, Volume I.2 - International Codes. Tech. rep., World Meteorological Organization (WMO) (2019). https://library.wmo.int/records/item/35625-manual-on-codes-volume-i-2-international-codes
21. World Meteorological Organization (WMO): Guide to the WMO Integrated Global Observing System. Tech. rep., World Meteorological Organization (WMO) (2021). https://library.wmo.int/records/item/55696-guide-to-the-wmo-integrated-global-observing-system#.YZTPeNDMKfA

Secure Usage of Containers in the HPC Environment

Barbara Krasovec[1,3]($\boxtimes$) and Teo Prica[2,3]

[1] Jozef Stefan Institute, Jamova 39, Ljubljana, Slovenia
barbara.krasovec@ijs.si
[2] IZUM, Presernova ulica 17, Maribor, Slovenia
[3] Slovenian National Supercomputing Network, Ljubljana, Slovenia

Abstract. Containers in High Performance Computing (HPC) provide a flexible and efficient alternative to traditional modular software. They allow researchers and engineers to package applications and their dependencies in a single file, ensuring portability, reproducibility and consistent performance across different computing environments. With the emergence of AI, industrial users and SMEs in HPC, containers have become even more convenient to use, as they can also meet the security requirements for isolating user workloads. As the HPC community embraces container technology, it is ushering in a new era of more efficient, repeatable and scalable research computing. However, adoption of containers in HPC has been relatively slow, largely due to misconceptions about how they work and concerns about potential performance degradation, energy inefficiency and security risks. This article looks at the security considerations when using containers and uses practical examples to show that, with the right configuration and tools, containers can be as secure as any other system application.

Keywords: Container · HPC · Security · Isolation · Threats

1 Introduction

Scientific software often has multiple and complex dependencies, some of which may be obsolete but are essential for a program to run. Building such software can be difficult, especially if a user needs to do it repeatedly in different environments. In the High Performance Computing (HPC) environment, scientific software is usually provided either as module files (e.g. built by the HPC owner using Spack or Easybuild) or as virtualised environments.

Virtualisation can be divided into two types. Infrastructure-level virtualisation abstracts the hardware and runs multiple operating systems on the same system. Container virtualisation is a form of OS-level virtualisation that abstracts the application layer. The host kernel allows multiple containers to run concurrently, sharing the host kernel but with different software stacks. Containers can run on either physical or virtual machines. They are suitable for a multi-tenant HPC environment because they allow multiple applications to run on

© The Author(s) 2025
A. Azab and T. Malkiewicz (Eds.): NeIC 2024, CCIS 2398, pp. 96–112, 2025.
https://doi.org/10.1007/978-3-031-86240-3_7

the same host without dependency conflicts. They also allow legacy code, which is common in academic scientific workflows, to run on the system. A further benefit is that they isolate applications with minimal or no performance overhead and provide portable and reproducible environments. Users can maintain their own containers. Compared to virtual machines, containers use resources more efficiently because they do not require their own operating system and libraries [24].

Containers are also a good fit for AI and industrial users on HPC. Unlike academic workflows, AI and industry workloads typically use the latest versions of software and libraries, as well as commercial software that can also be problematic to install on the HPC system. Industrial users are focused on increasing business value and competitiveness, they often rerun the same workloads, they strive for more standardisation, and they demand reproducibility, reusability, reliability and performance. Applications and calculations are highly optimised because inefficient workloads are costly. For such workloads, containers are an ideal solution. Containers also meet the stringent requirements of industrial HPC users in terms of process isolation, security and data lifecycle [16].

Despite the need for containers and the benefits they bring to HPC users, adoption of this solution in HPC has been relatively slow. This is due to resistance to change and inertia in the HPC community, which is accustomed to traditional software management methods; security and performance concerns [22], which often stem from a misunderstanding of how containers actually work; and compatibility or interoperability issues, as containers may not work well with some HPC components and services, such as schedulers, parallel file systems, and high-speed networking.

In this article, we will focus on the security aspects of using containers in the HPC environment and provide recommendations for secure container building, configuration, deployment and execution. The focus will be on Apptainer and Singularity as these are the most commonly used container solutions in the Slovenian National Supercomputing Network, with comparisons to Docker, Podman or other solutions where relevant.

2 Containers in the HPC Environment

HPC is a multi-tenant environment, designed to tackle complex computational tasks. The focus is on maximising performance, reducing computation time, and ensuring the accuracy and reliability of results. User access controls are applied based on the UID and GID, which are unique IDs assigned to users and groups. In addition, containers support additional isolation options, achieved through the use of namespaces, seccomp profiles and control groups (cgroups). Cgroups restrict the use of resources, while namespaces restrict what the user can see.

2.1 Container Isolation

Using namespaces is crucial for achieving process and resource isolation. A namespace wraps a global system resource in an abstraction that makes it appear to the processes within the namespace that they have their own isolated instance of the global resource. Linux kernel first introduced mount namespaces in version 2.4.29. Since then, additional options for isolation have been added:

- PID namespaces are used for process isolation,
- Network namespaces for managing/separating network interfaces,
- IPC namespaces for separating inter-process communication,
- Mount namespaces for managing/separating filesystem mount points,
- UTS namespaces for isolating kernel and version identifiers (mainly to set the hostname and domain name visible to the process),
- User namespaces for privilege isolation,
- Cgroup namespace for resource isolation.

To see which namespaces are used on the system `lsns` can be used:

```
# lsns
        NS TYPE    NPROCS    PID USER      COMMAND
4026531835 cgroup     219      1 root      /usr/lib/systemd/systemd --switched-root --system --deserialize 17
4026531836 pid        203      1 root      /usr/lib/systemd/systemd --switched-root --system --deserialize 17
4026531837 user       219      1 root      /usr/lib/systemd/systemd --switched-root --system --deserialize 17
4026531838 uts        203      1 root      /usr/lib/systemd/systemd --switched-root --system --deserialize 17
4026531839 ipc        203      1 root      /usr/lib/systemd/systemd --switched-root --system --deserialize 17
4026531840 mnt        196      1 root      /usr/lib/systemd/systemd --switched-root --system --deserialize 17
4026531860 mnt          1     39 root      kdevtmpfs
4026531992 net        203      1 root      /usr/lib/systemd/systemd --switched-root --system --deserialize 17
4026532241 mnt          1    626 root      /usr/lib/systemd/systemd-udevd
4026532243 mnt          1    706 root      /sbin/auditd
4026532244 mnt          1    739 chrony    /usr/sbin/chronyd
4026532245 mnt          1    737 root      /usr/sbin/irqbalance --foreground
4026532246 mnt          1 283643 root      /usr/sbin/NetworkManager --no-daemon
4026532250 mnt          5 550772 root      nginx: master process nginx -g daemon off;
4026532251 uts          5 550772 root      nginx: master process nginx -g daemon off;
4026532252 ipc          5 550772 root      nginx: master process nginx -g daemon off;
4026532253 pid          5 550772 root      nginx: master process nginx -g daemon off;
4026532255 net          5 550772 root      nginx: master process nginx -g daemon off;
4026532321 mnt          1   1116 root      /usr/sbin/rsyslogd -n
4026532478 mnt          8   2018 root      /bin/sh -c /code/run_uwsgi.sh
4026532479 uts          8   2018 root      /bin/sh -c /code/run_uwsgi.sh
4026532480 ipc          8   2018 root      /bin/sh -c /code/run_uwsgi.sh
4026532481 pid          8   2018 root      /bin/sh -c /code/run_uwsgi.sh
4026532483 net          8   2018 root      /bin/sh -c /code/run_uwsgi.sh
```

In the context of containers, mount namespaces are usually enabled by default to mount the file system into the container, other namespaces can be used but are optional. For the HPC environment, user and network namespaces are relevant and they are enabled by default in newer operating systems. User namespaces are a recommended option for most containerisation technologies available in HPC: Shifter, Podman, Apptainer, Singularity, LXD, CharlieCloud, Sarus, enroot and others. The isolation features included in Apptainer and Singularity are mount namespace by default, and optional namespaces are IPC, network, PID, UTS and user namespace. For container isolation we can also use other security controls, such as auditing rules, seccomp filters and Linux Kernel Capabilities.

User Namespaces have a broader application in enhancing system security. By creating isolated user spaces, they enable a regular (non-root) user to perform certain operations that would otherwise require superuser privileges, but being

restricted to that namespace [1]. In CentOS user namespaces were first introduced in the 3.8 Kernel (CentOS 7.6 and newer supported user namespaces),
but they were enabled by default in the CentOS 8 releases.

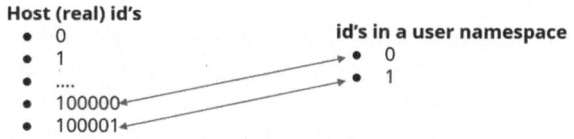

Fig. 1. User mapping when using user namespaces

By allowing unprivileged users to run containers or other processes requiring
root privileges within their own namespace, without affecting the host system
or other namespaces, they can reduce the attack surface and the risk of privilege escalation. One can limit the access and visibility of processes to system
resources and syscalls by using seccomp filters, capabilities and cgroups. This can
improve the security and efficiency of processes, and prevent them from interfering with each other or the host system (reducing the risk of noisy neighbours).
However, when unprivileged user namespaces are allowed on the system, a user
can use their own unprivileged installation of Singularity or any other runtime,
and they may also use FUSE mounts or tools like unshare or nsenter, which
means that they can bypass some of the system-wide settings, configured in the
container configuration file (including strict signature checking). Singularity and
Apptainer use a number of other kernel features to tackle these security concerns.
Containers are run as unprivileged users and can never enable privilege escalation on the host. The file system is mounted using the nosuid option, processes
are started with the PR_NO_NEW_PRIVS flag set [11]. To demonstrate that
a root user in the container has no privileges on the host, we will run a program
with some namespaces, unshare tool is used:

```
[barbara@host ~]$ whoami barbara
[barbara@host ~]$ unshare --fork --pid --mount-proc -U -r bash
[root@host ~]# whoami
root
[root@host ~]# less /var/log/messages
/var/log/messages: Permission denied
```

User namespaces can cause certain compatibility issues with some software or
frameworks that depend on the user's identity or permissions within the namespace, such as SSH or RSH. This software may not work correctly or securely
within user namespaces and may require additional configuration or modification. They also make process management more complex and costly, requiring
more steps and tools to set up, customise and track user namespaces. User
namespaces can expose processes to new risks or attacks. Users can exploit user
namespaces to bypass security checks, access forbidden resources or run malicious code. The latter can be prevented by not relying on check that can be

bypassed when using user namespaces, such as "if UID = 0", following the security guidelines and recommendations of developers and HPC administrators, using reliable and verified images or processes from trusted sources, scanning images or processes for vulnerabilities and malware before running them, and updating them regularly.

Sometimes, for example when installing additional software, it is necessary to have root privileges inside a container. This can be done by using unprivileged user namespaces or by using a setuid root utility. Setuid-root is a mode that allows a program to run with the privileges of its owner, which is usually the root user. Singularity and Apptainer can use a setuid-root utility, called starter-setuid, to run containers with root privileges without the need to be root or use sudo. Apptainer uses unprivileged user namespaces by default as of Apptainer 1.1. But which option is safer? A poorly coded setuid-root program can be an entry point for privilege escalation, code injection or other attacks. The vulnerability CVE-2019-11328[1] in Singularity 3.1.0 through 3.2.0-rc2 with setuid enabled allowed privilege escalation on the host due to insecure permissions that allowed a user to edit files in `/run/singularity/instances/sing/<user>/<instance>`. The manipulation of these files can change the behaviour of the starter-suid program when instances are joined. On the other hand, a potential danger of using user namespaces is that a process could bypass some security checks or restrictions on UID/GID, exploit the user namespace and gain access to some resources and do things that it would not normally be able to do. However, the use of user namespaces significantly reduces the attack surface and privilege escalation to root.

Network Namespaces are used to manage network interfaces. They are useful for ensuring that contained processes can bind the ports they need without interfering with each other. Traffic can be directed to a specific application. Interaction with network namespaces can be done by **nsenter** on the host.

```
# nsenter --target 248511 -n ip addr
1: lo: <LOOPBACK,UP,LOWER_UP> mtu 65536 qdisc noqueue state\
 UNKNOWN group default qlen 1000
    link/loopback 00:00:00:00:00:00 brd 00:00:00:00:00:00
    inet 127.0.0.1/8 scope host lo
       valid_lft forever preferred_lft forever
    inet6 ::1/128 scope host
       valid_lft forever preferred_lft forever
11: eth0@if12: <BROADCAST,MULTICAST,UP,LOWER_UP> mtu 1500 \
qdisc noqueue state UP group default
    link/ether 02:42:ac:11:00:02 brd ff:ff:ff:ff:ff:ff link-netnsid 0
    inet 172.17.0.2/16 brd 172.17.255.255 scope global eth0
       valid_lft forever preferred_lft forever
    inet6 2001:1470:ff8a:ba:0:242:ac11:2/64 scope global nodad
```

---

[1] https://cve.mitre.org/cgi-bin/cvename.cgi?name=CVE-2019-11328.

```
        valid_lft forever preferred_lft forever
    inet6 fe80::42:acff:fe11:2/64 scope link
        valid_lft forever preferred_lft forever
```

Since network namespaces are not required to successfully run a job in the HPC, they should be disabled [5]. Podman and Docker use network namespaces by default, but they can instead use the host's network. Charliecloud, Sarus, Apptainer do not require network namespaces by default. To disable network namespaces on the host:

```
echo "user.max_net_namespaces = 0" > /etc/sysctl.d/92-max_net_namespaces.conf
sysctl -p /etc/sysctl.d/92-max_net_namespaces.conf
```

Disabling network namespaces may have impact on some systemd services, for instance services that use PrivateNetwork feature, such as hostnamed or updatedb. This can be disabled by setting `PrivateNetwork=no`

```
cat /usr/lib/systemd/system/*.service| grep PrivateNetwork
```

Past vulnerabilities in namespaces have led people to believe that enabling them is a security threat, but the use of namespaces is a security feature designed to improve container isolation and security on the system. Most of the namespace vulnerabilities could not be exploited when network namespaces were disabled. There were numerous vulnerabilities in namespaces in the past few years, most of the vulnerabilities were related to the Netfilter.

Table 1. List of relevant high and critical vulnerabilities in namespaces

CVE-ID	Description
CVE-2021-22555[20]	A vulnerability in `net/netfilter/x_tables.c` allowed privilege escalation, public expoloits available.
CVE-2022-0185[21]	Kernel vulnerabilities, a local user was able to open a filesystem that does not support Filesystem Context API.
CVE-2022-32250[22]	A vulnerability in Netfilter.
CVE-2022-1015[23]	Vulnerability in Netfilter allowing privilege escalation.
CVE-2022-1016[24]	A vulnerability in Netfilter allowing privilege escalation.
CVE-2022-25636[25]	Memory access flaw in Netfilter subcomponent.
CVE-2023-0179[26]	A vulnerability in Netfilter allowed privilege escalation.
CVE-2023-32233[27]	Netfilter vulnerability, privilege escalation for local users with CAP_NET_ADMIN.
CVE-2024-1086[28]	Use-after-free vulnerability in nf_tables, enables privilege escalation.

With network namespaces disabled, only CVE-2023–32233 and CVE-2022-0185 could potentially be exploited in the HPC context. System administrators must monitor the kernel vulnerabilities and apply the fixes that could prevent privilege escalation or other exploits in the namespaces. Vulnerability management is crucial for all the software on the system, containers and their dependencies are no exception to this rule (Table 1).

Cgroups, short for control groups, are a way of controlling how much resources (such as CPU, memory, network bandwidth, etc.) a process or group of processes can use. In older versions of Linux, cgroups were not isolated from each other, which could lead to information leakage between containers. There are two versions of cgroups: v1 and v2. Cgroups v2 allows a process to belong to more than one cgroup, and allows more secure delegation of cgroups to containers. This means that users can give a container runtime, such as Docker or Podman, a part of the cgroup hierarchy to manage, without giving it full access to the whole cgroup system. This prevents the container runtime from affecting the cgroups of other processes or containers on the system, improving the security and isolation of the container. Cgroups v2 is supported by Linux kernel version 5.8 or later. Most modern Linux distributions, such as EL9, Ubuntu 22.04, Debian 11, Fedora 31 and newer, use cgroups v2 by default and allow unprivileged users to use the -apply-cgroups and other resource limiting flags of Apptainer without any additional configuration. Cgroups can be managed using the /sys pseudo file system. The resource usage of a specific container can be monitored there.

```
cat /sys/fs/cgroup/memory/docker/<container ID>/memory.max_usage_in_bytes
531476480
```

Linux Kernel Capabilities are a way of controlling the privileges of processes in a fine-grained way. They break down the privileges traditionally associated with the root user into about 40 different units that can be enabled or disabled for each process. This allows processes to perform specific operations that require elevated privileges without giving them full access to the system. The capabilities are a per-thread attribute, meaning that each thread in a process can have a different set of capabilities. As opposed to the use of setuid root, which is sensitive from a security perspective, administrators can grant permissions on a more granular basis with the use of Kernel capabilities. By default Apptainer and Singularity use CAP_SYS_ADMIN to perform various system administration tasks such as mounting file systems, creating device nodes and loading kernel modules, CAP_MKNOD to create special files with mknod, CAP_SETUID to perform arbitrary manipulations of process UIDs, CAP_SETGID to perform arbitrary manipulations of process GIDs, CAP_DAC_OVERRIDE to bypass file read, write and execute permissions checks, and CAP_CHOWN which allows changing the owner and group of files.

Granting additional privileges to users should be done carefully and only when really necessary [8, 12].

```
# apptainer capability add --user=barbara CAP_NET_RAW
# apptainer capability list barbara
barbara [user]: CAP_NET_RAW
```

To check the capabilities on the system **libcap-ng-utils** can be used

- **pscap** shows process-based capabilities,
- **filecap** shows file-based capabilities,

- `setcap` is used to add capabilities.

For example to check the process capabilities of Nginx docker container:

```
pscap| grep nginx
1951  1981  root      nginx        chown, dac_override, fowner, fsetid, \
kill, setgid, setuid, setpcap, net_bind_service, net_raw, \
sys_chroot, mknod, audit_write, setfcap
248492 248511 root    nginx        chown, dac_override, fowner, fsetid,\
 kill, setgid, setuid, setpcap, net_bind_service, net_raw, sys_chroot, \
 mknod, audit_write, setfcap
```

Seccomp Profiles. Seccomp stands for secure computing mode. Seccomp profiles are used to restrict container system calls from userspace to the kernel, and have been a feature of the Linux kernel since version 2.6.12.

For Singularity, the -`security` flag can be used by the root user to use security modules such as SELinux, AppArmor and seccomp inside the Singularity container. We can blacklist some folders, disable some system calls etc. Singularity/Apptainer do not enforce a default seccomp filter. Instead, they provide users with the option to apply custom seccomp filters as needed.

```
--security="seccomp:/usr/local/etc/singularity/seccomp-profiles/some_profile.json"
--security="apparmor:/usr/bin/man"
--security="selinux:context"
```

Docker has a default seccomp profile, but it can be disabled:

```
docker run --security-opt seccomp=unconfined
```

2.2   Rootless Containers

The idea of rootless containers refers to creating, running and managing a container as an unprivileged user, i.e. without gaining administrative privileges on the host. They offer several advantages, one of which is improved security. Even if the container engine, runtime or orchestrator is compromised, the attacker cannot gain root access to the host system. Rootless containers use user namespaces to map the user and group IDs inside the container to different ones on the host system. This allows processes inside the container to have root privileges (UID 0) without affecting the host system. Rootless containers also use other namespaces such as mount, network and PID to isolate the container's filesystem, network and processes from the host system.

In Apptainer and Singularity, the fakeroot feature enables rootless mode, allowing an unprivileged user to run a container as a fake root user by using user namespaces with user namespace UID/GID mapping. There are several ways to use the fakeroot feature. If the host is set up to map the current user via /etc/subuid and /etc/subgid mapping files, this method is used first. It is the most complete emulation, but requires administrator setup. This is also the method used by Podman. If this option is not available, the root-mapped user namespace is used, i.e. the root user id is mapped to the original unprivileged user. If the fakeroot command is available on the host, Apptainer will use it in addition to a root-mapped user namespace. Another option is to use fakeroot directly with setuid-root if user namespaces are not enabled [6].

2.3   Containers as Processes

A container is a running process controlled by the host kernel, it is isolated from the host and other processes. Processes on the running system can be found on /proc virtual filesystem and can be monitored and manipulated from the host system. What do these processes mean for security? Environment variables in containers often store secret information. A user with access to the underlying host can read the contents of the environ file inside /proc/PID/environ.

To see if a process is running as a container:

```
ps -ef --forest
root      550751      1  0 15:10 ?        00:00:00 /usr/bin/containerd-shim-runc-v2 -namespace moby -id \
b9315cdb9c9baf8901a21a3557daefbbebba9d71a7b868aca48de621a2e25dce -address /run/containerd/containerd.sock
root      550772 550751  0 15:10 ?        00:00:00  \_ nginx: master process nginx -g daemon off;
101       550818 550772  0 15:10 ?        00:00:00     \_ nginx: worker process
101       550819 550772  0 15:10 ?        00:00:00     \_ nginx: worker process
101       550820 550772  0 15:10 ?        00:00:00     \_ nginx: worker process
101       550821 550772  0 15:10 ?        00:00:00     \_ nginx: worker process
```

3   Containers and Security

The HPC community has a cautious approach to adopting containers, which is influenced by several security considerations, such as risk of privilege escalation, vulnerability management (containers are often outdated and software within the containers represents a security risk), compliance with security standards, complexity of management etc. These considerations require careful planning and implementation of security measures to ensure that container adoption does not compromise the HPC environment's integrity.

3.1   Images

Containers are built from layers of files called container images that define what runs in a container. The most common security threats are outdated or unnecessary software packages in the container image [21]. Other image-related security issues include the use of vulnerable base images, vulnerable dependencies in the image, hardened stored image or definition file, misconfigurations, insecure image storage etc. Images often come with complete operating system distributions, along with the necessary scientific software and its dependencies. Updates to these images are infrequent due to concerns that new software versions may introduce incompatibilities or unintentionally affect analytical results through unanticipated numerical changes. To facilitate reuse across different experiments, images are usually packed with more dependencies than are strictly necessary.

As a result, most images contain vulnerabilities, more than 30% of the images on the DockerHub [18].

To minimise the number of vulnerabilities in the image, unneeded packages should be removed to minimise the attack surface and the rest should be kept up to date to prevent them from becoming vulnerable. Images should be scanned for vulnerabilities. Several scanning tools are available, such as Clair, Anchore, Vuls, Trivy, OpenSCAP or similar. These scanners use vulnerability databases from specific OS distributions or they use Open Vulnerability and Assessment Language (OVAL). OVAL is an international information security community standard that promotes open and publicly available security content and includes support for a wide variety of operating systems and services. It is also important to note that software installed as files (not via package managers) is not scanned for vulnerabilities, so scientific code in the image should be verified and validated with other tools (ideally via a continuous integration process). It is better to build an image from scratch than to rely on third-party images. Publicly available images often do not provide definition files, the installed software in the image is unknown and the integrity of the image cannot be verified due to missing or expired signatures.

When writing the instructions to build container images, standards for securing the containers should be used, such as the CIS Benchmarks, NIST SP 800-190, the OWASP Top 10 and the OWASP Application Security Verification Standard [10]. When using a Dockerfile format to write a set of instruction, we can gain compatibility with different container solutions. Images should not contain any credentials or data; the data can be used, if required and with strict access controls, by bind-mounting external storage locations.

Images can also be encrypted and signed. For signing PGP keys or certificates can be used. Before a container is launched, the signature is verified - this ensures the integrity of the image. When creating and signing images, we need to make sure that the lifetime of the private key is limited, that the private key is stored securely and that it can be revoked in case of a compromise. Over time malicious users could gain access to the private key and try to insert malware in the image. To ensure optimal management of digital signatures, certificates should be periodically re-issued and containers re-signed. One option is also to use certificates without a key, which is a new model of image signing, introduced by SigStore project [3]. In automatic deployments, images can be signed by multiple keys, for each step in a CI/CD process. In setuid mode, we can make use of execution control list (ECL), which implements strict signature checking, at least one or more valid signatures should be found to run a container with this image.

There was a vulnerability in Singularity, CVE-2019-19724[29], exploiting it could compromise an image on the system. Due to incorrect file permissions on user configuration and cache directories, $HOME/.singularity was created with mode 777, potentially allowing anyone with login access to the host to modify container images and run malicious container images on behalf of affected users.

---

[29] https://nvd.nist.gov/vuln/detail/CVE-2019-19724.

Singularity versions between 3.3.0 and 3.5.1 are affected. The bug was fixed in version 3.5.2 [1].

Singularity and Apptainer support using containers in a single file, using SIF format. By holding an image in a single file, the container filesystem is immutable and can be signed. The signature is part of the image, so it is always possible to verify its integrity.

CI/CD Pipeline: most container vulnerabilities derive from using untrusted base images, misconfigurations, improper implementation of signatures in the code, and vulnerable software contained in the image. These threats can be mitigated by using automatic process to build containers.

Golden images are base images used as an entry point for creating custom software. The base image should come from a trusted and official repository, and the source code of the image should be available (do not use containers as a black box). Container-specific operating systems such as CoreOS should be used where possible, as they don't include package managers and have a minimal software stack.

Images should be scanned for vulnerabilities; if images are deployed in an automated way, scanning should be part of the CI process, using static application security testing (SAST) or dynamic application security testing (DAST). Images should be signed and the signature verified before the container is run. However, not all container solutions support signing containers, e.g. enroot doesn't. Singularity and Apptainer do support signing. With the implementation of this feature, some vulnerabilities related to signature signing and verification were discovered in 2020, namely CVE-2020-13845[30] with execution control list privilege bypass, CVE-2020-13846[31] where signature verification did not display a missing signature warning, and CVE-2020-13847[32] where signing or verifying an image did not include metadata. All three issues have been fixed in Singularity version 3.6.0.

Storage. Where possible, images should be stored in a private registry and made available to users. By using private registries, we can control access to the images and implement other security features. Images are typically created using CI/CD pipelines, a shift-left approach to deployment should be adopted. A shift-left approach focuses on security measures at all stages of the development cycle to ensure security before the software is released or before the container image is placed in the registry. In other words, testing is shifted to the left on the project timeline. CI/CD process includes style testing, code quality and performance testing, conformance testing, SAST testing, where the source code is scanned for vulnerabilities, DAST testing, where the application is exposed to various attacks, secrets detection, where the code is scanned for secrets, dependency

---

[30] https://github.com/hpcng/singularity/security/advisories/GHSA-pmfr-63c2-jr5c.

[31] https://github.com/hpcng/singularity/security/advisories/GHSA-6w7g-p4jh-rf92.

[32] https://github.com/hpcng/singularity/security/advisories/GHSA-m7j2-9565-4h9v.

scanning for security issues, and so on. This process can prevent many security threats, such as adding malware to the image. To increase the level of security, containers can also be encrypted, but some containers do not support encryption, e.g. enroot.

3.2   Containers at Runtime

Securing containers at runtime means detecting and preventing unexpected behaviour from applications that run within the container. The attack surface should be minimised by host hardening and container isolation, using namespaces, cgroups, seccomp profiles, using SElinux, Apparmor, auditing and configuring Linux kernel features [17]. Container escapes that enable the adversaries to run malicious code on the host from inside an isolated container are not rare, but are not well studied. These attacks are possible in case of container misconfigurations, in case of kernel vulnerabilities or container runtime vulnerabilities [21], such as CVE-2019-5736[33] or CVE-2019-19921[34].

4   Security Threats

Organisations that value a strong security posture must be able to address common container security issues, although there are multiple security controls that should be in place based on the threat being mitigated. In this chapter we will describe some of the most common threats in containerised environments:

- Software vulnerabilities: vulnerabilities on the host system and in the image can represent a serious threat. Images and systems should always be checked for outdated packages and software vulnerabilities. The prevention was discussed already, vulnerability scanning should be in place, the system and the image updated and patched, unnecessary packages removed.
- Build image attacks: when an attacker can modify the way the image is built, by exploiting the build host, adddding crypto mining or malware in the image, inserting some other malicious code in the image, accessing build secrets, exposed build secrets.
- Shared host attack - container escape and attack on the containers on the same host.
- Stealing exposed credentials.
- Container image attack: container consists of executable software (search for common malware strings: coinhive, base64_decode) [14].
- Data leak, access to private data in the container: use a secure external location for the data.
- Insecure networking.
- Root kernel attack: malicious container overwrites binary of runc and gains root access: see RunC vulnerabilities, such as CVE-2019-5736[35].

---

[33] https://cve.mitre.org/cgi-bin/cvename.cgi?name=CVE-2019-5736.

[34] https://cve.mitre.org/cgi-bin/cvename.cgi?name=CVE-2019-19921.

[35] https://cve.mitre.org/cgi-bin/cvename.cgi?name=CVE-2019-5736.

Other threats include privilege escalation, information leak, external attackers, insider threats, inadvent actors inside, problematic application processes, exposed secrets, insecure networking, inserted malicious code, human error, and inproper permissions or access controls [25].

5   Container Security (Mis)conceptions

Software security has become a major concern, not only in the HPC environment, but in general. One aspect is the need for more standardisation and transparency in software development, and improving the security and integrity of the software supply chain, especially for commercial software (e.g. SolarWinds case) [20]. The vulnerabilities in the Log4j framework highlighted another issue, namely that security cannot be taken for granted, even when using trusted libraries that are widely used as dependencies in the development cycle. The authenticity and integrity of software should be verified before it is used. But how do you ensure the integrity of the software? In the case of containers, the processes of signing them with X.509 certificates or PGP keys is a straightforward solution, a bigger issue is secret/key management and their lifecycle. We have mentioned the importance of container image signatures in previous chapters. Signature verification also includes detecting container changes and possible software vulnerabilities. The fact that a container has not been updated for a while virtually guarantees that it contains a variety of CVEs and opportunities for attackers. Not checking the signatures of the images is one problem, another is that the signature has expired and we use the container anyway.

With all the isolation options that are available in the kernel, containers are a valid solution for secure application deployment and usage, but in the HPC environment they are not as widely used as they could be. This is partly due to some misconceptions about their use. Some people believe that containers cause performance degradation because they are perceived as virtual machines. Others believe that containers introduce additional security threats, due to unhandled vulnerability issues in the software installed. To argue the first, several studies have shown that containerisation has no or minimal effect on performance, so this concern can be dismissed [19,23]. The NIST standard states that *"Security is a major concern when using containers in HPC environments. Containers have large attack surfaces due to the different underlying images, each of which may have vulnerabilities. In addition, securing the host is not enough to ensure protection. Container permissions and proper isolation are also required. Finally, containers can be difficult to monitor due to their dynamic nature"* [24]. However, containers are just like any other application on the system. They need to be kept up to date and verified, and they need to be configured correctly. Monitoring containers in the HPC is challenging, as any other workload, but not impossible. It is difficult to detect anomalies or malicious jobs in the user workload because it is constantly changing. Nodes are often not monitored extensively because they are numerous and their monitoring is minimised to detect the critical issues. Administrators make use of so called healthcheck software, that performs automated checks on the node without using an external monitoring system. Auditing rules should be set on the host to detect anomalies and report them, such as detecting tasks executed by root, failed logins, modification of critical files or binaries etc.

With the focus on industry and AI users on the HPC systems, HPC environment will have to adopt some practices from the cloud world, especially the orchestrators to manage multiple containers, API support and CI/CD support for the users' workloads. Interoperability and integration of HPC, cloud and big data will also bring other security considerations, especially for secure usage of these tools and providing secure data channels for the workloads [13,15,27].

5.1 Containers on HPC Vega

HPC Vega [7] is a petascale EuroHPC supercomputer that provides computing resources to the Slovenian and European scientific community. SingularityPRO is installed cluster-wide, in our virtualised environment we also support Apptainer. SingularityPRO is a licensed version, which means that we are informed about upcoming releases and potential security issues before they are publicly announced. The Singularity installation supports setuid mode with user namespaces enabled, network namespaces disabled. Fakeroot permissions are added manually on request by adding the user to /etc/subuid and /etc/subgid. We believe that using this feature does not introduce any additional security risks, potential zero day vulnerabilities in the singularity or namespaces could lead to privilege escalation, but the likelihood is low as we have proper vulnerability management in place and multi-factor authentication for users accessing the cluster. We encourage users to use base images from trusted sources when building their own container environment, and we provide template Singularity definition files that include interconnect settings and other cluster specifics to facilitate the process. We also maintain containers for specific applications, particularly machine learning frameworks, which require recent versions of software that are not always easy to install. Users often use images from the Docker Hub [4] and NVIDIA NGC Catalog [9].

To ensure proper isolation of containers, we've enabled user namespaces, disabled network namespaces, and disabled control groups, as jobs run through the Slurm batch system, which has its own cgroup namespace implementation. Cgroups are configured for memory, CPU and devices. Cgroups v1 is used. A transition to Cgroups v2 is already planned, as this will allow additional granularity in resource handling [26]. Additional Linux features are not enabled.

5.2 Future Development on HPC Vega

HPC Vega uses a CernVM-FS (CVMFS) [2] repository to make scientific software available to users. Currently we provide software in the form of modules, we also have Singularity Sandbox environments available, specifically for the HEP workload. These environments are currently maintained by the HEP research groups, but are also monitored by a central Pakiti service which reports if there are any high or critical vulnerabilities in the containerised environment. In the future, we will restrict the usage of images on the cluster. We are implementing a shift-left approach to security by configuring an automated CI/CD workflow for building images. This workflow will include security testing and vulnerability management. A pull request from a user wil be required to push their Dockerfile to the repository. Container will be built using gitlab-runner, tested for code discrepancies, linted, scanned for vulnerabilities, etc. If the CI process is successful, the image will be published to the local Harbor image repository. A snapshot of this image will be done by CVMFS DUCC (Daemon that Unpacks Container Images into CernVM-FS) and published in the form of a sandbox environment on the local CVMFS repository.

5.3   Conclusion

In this article, we have discussed the security aspects of using containers and the many ways in which container workloads can be isolated and secured. We have described some of the major historical vulnerabilities relevant to this area and shown that many of them can be mitigated with proper configuration, deployment and vulnerability management, making containers as problematic from a security perspective as any other application. But with the benefits they bring to HPC users, enabling reproducibility, portability and support for both legacy and newer software libraries, they are an essential tool for deploying software in HPC.

Acknowledgements. The authors would like to thank Alja Prah, Daniel Kouril, Jan Jona Javorsek and Dejan Lesjak for their constructive comments.

References

1. Singularity version 3.5.2 release notes (2019). https://github.com/sylabs/singularity/releases/tag/v3.5.2
2. CernVM-FS. https://cvmfs.readthedocs.io/en/stable/ (2023)
3. sigstore. https://www.sigstore.dev/ (2023)
4. Docker Hub. https://hub.docker.com/ (2024)
5. EGI CSIRT, Linux namespaces and containers (2024). https://csirt.egi.eu/2022/10/19/linux-namespaces-and-containers/
6. Fakeroot feature (2024). https://apptainer.org/docs/user/main/fakeroot.html
7. HPC Vega Documentation. https://doc.vega.izum.si/ (2024)
8. Linux Kernel Capabilities (2024). https://man7.org/linux/man-pages/man7/capabilities.7.html
9. NVIDIA NGC Catalog. https://catalog.ngc.nvidia.com/ (2024)
10. OWASP Application Security Verification Standard (2024). https://owasp.org/www-project-top-ten/
11. Security in singularityce (2024). https://docs.sylabs.io/guides/3.11/user-guide/security.html

12. Singularity capabilities (2024). https://docs.sylabs.io/guides/3.0/admin-guide/configfiles.html
13. Abdelmassih, C.: Container Orchestration in Security Demanding Environments at the Swedish Police Authority
14. Ashutosh, C., Rajewar, R.: Cryptojacking Docker Images to Mine for Monero (2020). https://unit42.paloaltonetworks.com/cryptojacking-docker-images-for-mining-monero/
15. Beltre, A.M., Saha, P., Govindaraju, M., Younge, A., Grant, R.E.: Enabling HPC workloads on cloud infrastructure using kubernetes container orchestration mechanisms. In: 2019 IEEE/ACM International Workshop on Containers and New Orchestration Paradigms for Isolated Environments in HPC (CANOPIE-HPC), pp. 11–20. IEEE, Denver, CO, USA (Nov 2019). https://doi.org/10.1109/CANOPIE-HPC49598.2019.00007, https://ieeexplore.ieee.org/document/8950981/
16. Brayford, D., Vallecorsa, S., Atanasov, A., Baruffa, F., Riviera, W.: Deploying AI frameworks on secure HPC systems with containers. In: 2019 IEEE High Performance Extreme Computing Conference (HPEC), pp. 1–6. IEEE, Waltham, MA, USA (Sep 2019). https://doi.org/10.1109/HPEC.2019.8916576, https://ieeexplore.ieee.org/document/8916576/
17. Gantikow, H., Reich, C., Knahl, M., Clarke, N.: providing security in container-based hpc runtime environments. In: Taufer, M., Mohr, B., Kunkel, J.M. (eds.) High Performance Computing, vol. 9945, pp. 685–695. Springer International Publishing, Cham (2016). https://doi.org/10.1007/978-3-319-46079-6_48, http://link.springer.com/10.1007/978-3-319-46079-6_48, series Title: Lecture Notes in Computer Science
18. Kaur, B., Dugré, M., Hanna, A., Glatard, T.: An analysis of security vulnerabilities in container images for scientific data analysis. GigaScience 10(6), giab025 (Jun 2021).https://doi.org/10.1093/gigascience/giab025, https://academic.oup.com/gigascience/article/doi/10.1093/gigascience/giab025/6291571
19. Liu, P., Guitart, J.: Performance characterization of containerization for HPC workloads on InfiniBand clusters: an empirical study. Clust. Comput. 25(2), 847–868 (Apr 2022) 10.1007/s10586-021-03460-8, https://link.springer.com/10.1007/s10586-021-03460-8
20. Martínez, J., Durán, J.M.: Software supply chain attacks, a threat to global cybersecurity: solarwinds' case study. Int. J. Safety Secur. Eng. 11(5), 537–545 (Oct 2021). https://doi.org/10.18280/ijsse.110505, https://www.iieta.org/journals/ijsse/paper/10.18280/ijsse.110505
21. Reeves, M., Tian, D.J., Bianchi, A., Celik, Z.B.: Towards improving container security by preventing runtime escapes. In: 2021 IEEE Secure Development Conference (SecDev), pp. 38–46. IEEE, Atlanta, GA, USA (Oct 2021). https://doi.org/10.1109/SecDev51306.2021.00022, https://ieeexplore.ieee.org/document/9652631/
22. Rudyy, O., Garcia-Gasulla, M., Mantovani, F., Santiago, A., Sirvent, R., Vazquez, M.: Containers in HPC: a scalability and portability study in production biological simulations. In: 2019 IEEE International Parallel and Distributed Processing Symposium (IPDPS), pp. 567–577. IEEE, Rio de Janeiro, Brazil (May 2019). https://doi.org/10.1109/IPDPS.2019.00066, https://ieeexplore.ieee.org/document/8820966/
23. Ruiz, C., et al., (eds.): Euro-Par 2015: Parallel Processing Workshops, vol. 9523, pp. 813–824. Springer International Publishing, Cham (2015). https://doi.org/10.1007/978-3-319-27308-2_65, http://link.springer.com/10.1007/978-3-319-27308-2_65, series Title: Lecture Notes in Computer Science

24. Souppaya, M., Morello, J., Scarfone, K.: Application container security guide. Tech. Rep. NIST SP 800-190, National Institute of Standards and Technology, Gaithersburg, MD (Sep 2017). https://doi.org/10.6028/NIST.SP.800-190, https://nvlpubs.nist.gov/nistpubs/SpecialPublications/NIST.SP.800-190.pdf
25. Sultan, S., Ahmad, I., Dimitriou, T.: Container security: issues, challenges, and the road ahead. IEEE Access **7**, 52976–52996 (2019). https://doi.org/10.1109/ACCESS.2019.2911732, https://ieeexplore.ieee.org/document/8693491/
26. Yoo, A.B., Jette, M.A., Grondona, M.: Slurm: Simple linux utility for resource management. In: Workshop on job scheduling strategies for parallel processing, pp. 44–60. Springer (2003)
27. Zhou, N., Georgiou, Y., Zhong, L., Zhou, H., Pospieszny, M.: Container orchestration on HPC systems. In: 2020 IEEE 13th International Conference on Cloud Computing (CLOUD), pp. 34–36. IEEE, Beijing, China (Oct 2020). https://doi.org/10.1109/CLOUD49709.2020.00017, https://ieeexplore.ieee.org/document/9284299/

NordIQuEst: The Nordic–Estonian Quantum Computing e–Infrastructure Quest

Costantino Carugno[1,4(✉)], Jake Muff[1,2], Mikael P. Johansson[2], Sven Karlsson[3], and Alberto Lanzanova[2]

[1] VTT Technical Research Center of Finland, Espoo, Finland
costantino.carugno@vtt.fi
[2] CSC – IT Center for Science, Espoo, Finland
[3] Technical University of Denmark, Kongens Lyngby, Denmark
[4] DEIB, Politecnico di Milano, Milan, Italy

Abstract. This paper presents the Nordic–Estonian Quantum Computing e–Infrastructure Quest – NordIQuEst – an international collaboration of scientific and academic organizations from Denmark, Estonia, Finland, Norway, and Sweden, working together to develop a hybrid High–Performance and Quantum Computing (HPC+QC) infrastructure. The project leverages existing and upcoming classical high–performance computing and quantum computing systems, facilitating the development of interconnected systems. Our effort pioneers a forward–looking architecture for both hardware and software capabilities, representing an early–stage development in hybrid computing infrastructure. Here, we detail the outline of the initiative, summarizing the progress since the project outset, and describing the framework established. Moreover, we identify the crucial challenges encountered, and potential strategies employed to address them.

Keywords: Quantum Computing · HPC · Hybrid Computing

1 Introduction

Over the years, HPC systems have served as the foundation for conducting demanding calculations and simulations. Increasingly powerful supercomputers have been deployed with elaborate software stacks and dedicated resource management [15]. A supercomputer is realized through a computer cluster, equipped with a specialized set of software and interconnected hardware that enables massively parallel and distributed computing, leveraging the power of many individual computing units. Recently, QC has emerged as an alternative computational paradigm, advocating the use of quantum devices in order to carry out specialized computation in a radically different way, by leveraging quantum effects [7]. The availability of first-generation quantum computers, named "Noisy Intermediate-Scale Quantum" (NISQ) devices [18], has allowed researchers to

© The Author(s) 2025
A. Azab and T. Malkiewicz (Eds.): NeIC 2024, CCIS 2398, pp. 113–127, 2025.
https://doi.org/10.1007/978-3-031-86240-3_8

experiment with a variety of quantum algorithms under practical, although limited, conditions. Presently, these devices, characterized by their ability to handle from a few to a few thousand "noisy" qubits, face the issue of preserving information encoded in a quantum system for the required time necessary to carry out computationally difficult tasks [21]. However, current improvements in technology and future efforts towards developing "Fault–Tolerant Quantum Computers" (FTQC) [20], hold the potential to overcome this issue, and solve problems that are currently intractable for classical supercomputers alone [6]. From this, efforts towards integration of HPC systems with QC have sprung forth, with the aim of ultimately addressing uniquely complex scientific and engineering problems.

The NordIQuEst project[1] represents a pioneering initiative to leverage the complementary strengths of both HPC and QC computing systems, blending these two different forms of computing into a hybrid HPC+QC infrastructure [11]. This collaborative effort, promoted by the Nordic e–Infrastructure Collaboration NeIC[2], and brought forward by various organizations within the Nordic nations – DTU[3] (Denmark), VTT[4] and CSC[5] (Finland), Simula[6] and SINTEF[7] (Norway), Chalmers–WACQT[8] (Sweden), and University of Tartu[9] (Estonia) – aims to set up a prototypical infrastructure, incorporating HPC and QC devices, and specialized software. Integrating quantum computing into the existing framework of HPC systems introduces several challenges, including the need for compatible software environments, efficient management of resources, and robust access control [12]. NordIQuEst tackles these issues by developing a cohesive ecosystem that is approachable and reliable, in order to enable researchers to execute smoothly on both real QC devices and the hybrid system. In addition to preparing the infrastructure, NordIQuEst has organized several events and workshops aimed at disseminating knowledge and educating experimenters on HPC and QC topics[10][11][12][13]. The project reflects the Nordic tradition of collaboration in innovation and scientific inquiry, and seeks to establish a blueprint for the integration of quantum and classical computing resources.

This paper outlines the foundational principles of HPC and QC, examines the potential for their combination, and describes the implementation of the NordIQuEst platform. It addresses the technical, operational, and conceptual challenges of merging these computing paradigms, and shares the project's

---

[1] https://nordiquest.net/.

[2] https://neic.no/.

[3] https://www.dtu.dk/english.

[4] https://www.vttresearch.com/en.

[5] https://www.csc.fi/.

[6] https://www.simula.no/.

[7] https://www.sintef.no/en/.

[8] https://www.chalmers.se/en/centres/wacqt/.

[9] https://ut.ee/en/home.

[10] NordIQuEst workshop 2022.

[11] Quantum Computing Norway 2022.

[12] Quantum Autumn School 2023.

[13] Norway Young Ambassadors Program.

approach and vision for navigating current and upcoming obstacles. This work lays the foundation for further advancement of integrated HPC+QC computing technologies, upon which future projects may leverage and expand upon.

2   HPC and QC

HPC systems are advanced computing systems that leverage the aggregation and orchestration of computational resources, in order to tackle problems that demand vast amounts of computing power. Such infrastructures are constituted by the combined integration of several technologies, of which a supercomputer, in the form of a computer cluster, stands as the foundational element. A cluster is a physical infrastructure that groups many computing nodes – individual computing resources – that serve as a unified computational resource. The nodes are connected through a high–speed network and specialized interconnect hardware, specifically engineered to meet computational demands, leveraging the combined capabilities of one or more CPUs, GPUs, or other dedicated hardware. In addition, supercomputers are equipped with the memory and the storage necessary to accommodate the data. The supercomputer is managed on–site and is accessed remotely by users, that carry out intensive calculations using distributed parallel computing, a computational paradigm that consists in dividing the problem into smaller tasks, distributing and executing them concurrently across the nodes. Moreover, a group of tasks can be collected and processed simultaneously in batches, further reducing processing time. The efficiency of HPC systems heavily relies on the software and algorithms used, as optimizing execution for parallel processing is critical for maximizing performance. Workload management and job scheduling software is utilized, to allocate the computational resources needed without hindering overall system performance. Furthermore, a dedicated system to manage user identity and permission levels is essential, in order to properly assign computational quotas, and impose security restrictions, averting misuse of the system. Lastly, handling large volumes of data necessitates an advanced storage systems, such as a parallel file system, designed for big data storage, fast data retrieval, and high reliability.

QC represents a paradigm shift in our approach to computational problems, introducing principles that fundamentally diverge from classical computing. [23] At its core, quantum computing encodes information in a quantum system, and leverages phenomena from quantum mechanics – superposition, interference, and entanglement – to carry out the calculation. The fundamental unit of quantum information is the qubit, analogous to the bit in classical computing. Unlike a classical bit, which can take either one of two values – 0 or 1 – a qubit can exist in a superposition of both states, allowing, in principle, a quantum computer with n qubits to represent and process 2^n possible qubit states simultaneously. In classical computers, the information is represented as a string of n bits – a bitstring – which is inserted in a processor register, and after certain logic operations are performed, the resulting bitstring is obtained. In quantum computing, information is encoded in n qubits – a quantum state – symbolized as a quantum register but realized in a physical system. The qubits are then manipulated using quantum gates, i.e., basic logic operations, until they reach a final state composed of a superposition of the qubits states. By measuring the qubits, the final state collapses to a classical state, a bitstring. However, as the final state is the result of a superposition of the qubit states, a single measurement yields only one possible bitstring, obtained according to its specific probability, dictated by quantum mechanics. In order to reconstruct the probability distribution of all the outcoming bitstrings in the final state, the measurement process needs to be repeated several times, which is often referred to as "shots". In fact, quantum interference, the ability of quantum states to amplify or cancel each other, plays a critical role during quantum computation, guiding the state towards the correct solutions by enhancing probabilities associated with desired outcomes, while suppressing the others. Similarly to how a series of logical gates designed to perform specific operations are assembled into circuits, a series of quantum gates are assembled in quantum circuits, which are unitary transformations that enable superposition and entanglement, a phenomenon where the state of one qubit is correlated with the state of another, such that the measurement of one immediately defines the state of the other.

A given problem is thus solved by QC in a radically different way, with a quantum algorithm that pre–processes the data, encodes it in a qubit register, executes a quantum circuits several times, and post–processes the outcome. A diverse set of quantum algorithms has been conceived over the years, aimed at addressing issues that are difficult to solve using classical methods. It is important to note that quantum computers and algorithms will not replace classical computers and algorithms. In fact, one should think of QC devices as more closely resembling hardware accelerators, special–purpose solvers, rather than entire classical computers replacements. The concept of having dedicated hardware to perform specific demanding computational operations is not new to computer science. Initially conceived to advance computer graphics and image processing, Graphics Processing Units (GPUs) have proven useful in a diverse set of applications, from training neural networks to mining cryptocurrency, due to their capacity of handling intensive matrix calculations, and they are now a common component of HPC systems. Furthermore, while Application–Specific Integrated Circuits (ASICs) are custom–designed circuits, used in HPC for tasks that require high efficiency and performance, Field–Programmable Gate Arrays (FPGAs) have a versatile internal logic, and are utilized in HPC for their flexibility, and their ability to be reprogrammed for different scopes. Following a similar logic, the term Quantum Processing Unit (QPU) was recently established to identify the computational capability offered by QC devices.

The advantages of integrating HPC and QC into a hybrid HPC+QC infrastructure come in many forms. The following points delineate some of the prevalent use–cases of an HPC+QC system:

(a) **Hybrid classical–quantum algorithms**: Quantum circuits often feature a modular structure, with parameterized gates that are modifiable through classical inputs. Variational Quantum Algorithms (VQA) – such as the Variational Quantum Eigensolver (VQE) for chemistry problems [17], and the

Quantum Approximate Optimization Algorithms (QAOA) for combinatorial optimization problems [8] – require the interplay with classical routines to be tuned to the specific objective. In this scenario, classical computers can run parallel parameter optimization using one or more optimizers, whilst quantum computers are used to prepare quantum states and measure observables, which are then fed back into the classical optimization loop running on a classical resource. Furthermore, as the relationship between the classical parameters and the cost function is non–linear, the optimization landscape of VQA problems can be highly complex with many local minima, and, even in the case of a single optimizer, providing convergence to acceptable values can be challenging for ordinary computers.

(b) **QC supports HPC**: In the realm of machine learning, several hybrid algorithms have been developed that are mainly classical but could benefit from having a quantum component. A Quantum Support Vector Machine (QSVM) [19] is a classifier that utilizes a quantum kernel to divide the dataset. In such case, having an HPC infrastructure available allows high–performance pre–processing of large datasets and faster execution time, and to promptly compare the results obtained using a classical kernel. In Quantum Generative Adversarial Networks (QGANs) [5], whilst HPC trains the discriminator and optimize the overall network, the quantum computer runs the quantum generator, exploring the high–dimensional space in order to generate new data samples that can be used in the discriminator. In a similar way, for deep learning, in Quantum Convolutional Neural Networks (QCNNs) [4], whilst HPC performs the heavy computational tasks associated with training the classical layers, the quantum computer takes care of implementing the quantum layers.

(c) **HPC supports QC**: In addition to the classical processing required for variational hybrid algorithms, HPC resources are crucial for extracting the most out of quantum computers through pre– and post–processing of data. Pre–processing includes optimal compiling and transpiling of circuits, and increasingly, creating the quantum algorithms in the first place. [14] Efficient compiling and transpiling is necessary in order to make the final executable circuit sufficiently shallow for NISQ devices to finish before decoherence. With increasing qubit count, qubit routing and gate optimization becomes a computationally hard problem, requiring supercomputing resources. Post–processing includes error mitigation measures, in effect, a process of enhancing the signal–to–noise ratio. Also this task becomes increasingly resource–intensive as qubit number grows. In order to avoid losing the computational advantage of a conceptually efficient quantum algorithm, it is important that the pre– and post–processing tasks can be performed efficiently, that is, with low scaling with respect to qubit count. Here, machine learning techniques are expected to play a decisive role in suppressing the scaling of classical processing methods. [9]

(d) **HPC simulates QC**: HPC is crucial for benchmarking quantum computers, in assessing the current capabilities and scalability of quantum algorithms, and predicting their future potential. On a classical computer, the exact simulation of a quantum circuit of n qubits requires storing the complex amplitudes for 2^n possible states. As the memory necessary to accommodate these values grows exponentially with circuit size, the evaluation of circuits with around 25 and more, quickly becomes unfeasible for an ordinary machine. Instead, a supercomputer can take advantage of a distributed memory approach, distributing the complex values to different nodes, thereby pushing the boundaries of classical simulation around the 50 qubits mark, thus enabling larger scalability analysis. Conversely, approximate evaluations of locally–entangled quantum circuits can be classically performed using tensor network methods, which encode and compress a circuit, alleviating the aforementioned memory constraint. Depending on the specific circuit and the tensor structure chosen, it is sometimes possible to approximate circuits with a large number of qubits, even hundreds or thousands, provided that the entanglement is low enough [16]. However, in such cases, in order to reach the final result, several tensor contractions need to be performed and an efficient contraction path needs to be found, shifting the computational burden from memory to processors [10]. In conclusion, HPC is still indispensable in providing the necessary computational power for these simulations [22].

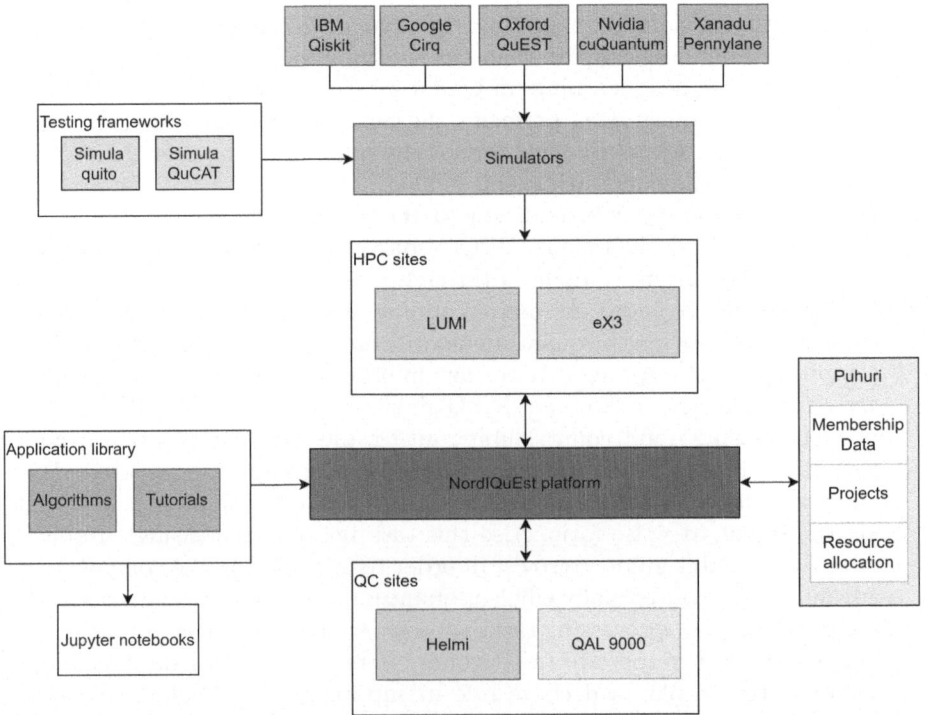

Fig. 1. A diagram of the NordIQuEst infrastructure, showing how the different components are connected to each another. The colors of the HPC and QC sites boxes reflect the location of the hardware: Finland (light blue), Norway (red), Sweden (yellow). (Color figure online)

3 NordIQuEst Infrastructure

3.1 Overview

The NordIQuEst platform acts as the central actor to manage the interaction between HPC systems and QC devices, as shown in Fig. 1. The infrastructure is realized through the integration of several hardware and software components, which NordIQuEst enables to seamlessly interface:

(a) **HPC sites**: The pan–European EuroHPC LUMI[14], is presently the fastest supercomputer in Europe and number five on the global TOP500 list[15]. LUMI consists of 2048 LUMI–C CPU nodes, each with two AMD EPYC 7763 CPUs with 64 cores, 2978 LUMI–G nodes, each with four AMD MI250x GPUs, and additional specialized computing partititons. All LUMI compute nodes use the HPE Cray Slingshot–11 200 Gbps network interconnect. LUMI has a sustained computing power of 380 PFLOPS, and is equipped with 1.75 PB of RAM. It is hosted by CSC – IT Centre for Science (Finland), and is the principal HPC resource of NordIQuEst. In addition, the eX3[16] project, coordinated by Simula (Norway), provides several different computational clusters for Norwegian researchers. HPC sites are equipped with the SLURM workload manager[17], to schedule jobs and dispatch them to QC sites.

(b) **QC sites**: Helmi[18], a 5–qubit superconducting quantum computer, hosted at VTT Research Center (Finland), is the main QC site for NordIQuEst. The qubits are arranged in a star–topology and have fidelities around 99.7% for single–qubit gates, 95% for two–qubit gates, and 95% for readout, with T2 times in the range of $10–20\mu s$. Of a similar superconducting technology, QAL 9000[19], is currently a 25–qubit testing chip, with planar grid connectivity map, fabricated and maintained by Chalmers–WAQCT (Sweden). Both QC sites are planning to upgrade to a 40–50 qubits device in the upcoming years.

(c) **User management**: Puhuri[20] is a cloud service project, funded by NeIC, which provides identity management, project management and resource allocation across Europe. It is integrated into both HPC and QC systems.

(d) **Programming frameworks**: Several QC simulators are available, in order to accommodate different needs. IBM Qiskit[21] and Google Cirq[22] are popular well–supported and well–equipped quantum programming frameworks. Oxford QuEST [13] is specially made for running on multi–core and multi–node clusters, in addition to supporting distributed memory. Nvidia cuQuantum [2] is designed for running on GPUs and performing advanced tensor

---

[14] https://www.lumi-supercomputer.eu/.
[15] https://www.top500.org/.
[16] https://www.ex3.simula.no/.
[17] https://slurm.schedmd.com/.
[18] https://vttresearch.github.io/quantum-computer-documentation/helmi/.
[19] https://www.qal9000.se/.
[20] https://puhuri.io/.
[21] https://www.ibm.com/quantum/qiskit.
[22] https://quantumai.google/cirq.

networks calculations. Xanadu Pennylane [3] contains several tools to run quantum machine learning algorithms and is suited for running large scale hybrid algorithms on supercomputers [1].

(e) **Testing frameworks**: In addition to the simulators, customized Python software for circuit testing is made available and maintained by Simula: quito[23], an automatic test coverage generator tool, and QuCAT[24], a quantum circuit analyzer tool.

(f) **Application library**: A quantum algorithm library, written in Python, is provided, in order to test and evaluate the capabilities of the hybrid computing framework. Introductory tutorials and additional educational resources are publicly available for new researchers and are easily executable through Jupyter notebooks.

3.2 Operational Workflow

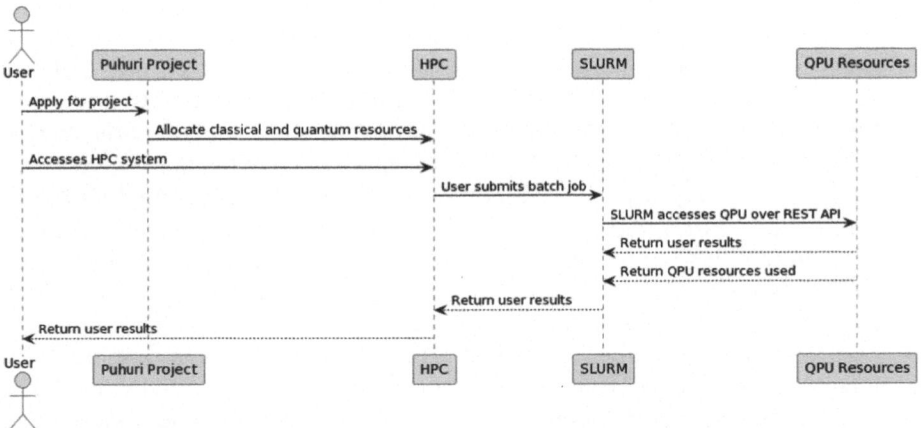

Fig. 2. A diagram of a typical user experience on the NordIQuEst platform.

The operational workflow designed within the NordIQuEst infrastructure, displayed in Fig. 2, is constructed to be intuitive, and to allow users to easily submit computational jobs to either HPC clusters, QC devices, or a combination of both. Initially, the user applies for a project through Puhuri. Upon approval, Puhuri allocates the necessary classical and quantum resource quotas. The user then accesses the HPC system directly. For job submission, the user sends single or a batch of jobs to the HPC, which are handled by the SLURM workload manager. SLURM manages the job execution, based on the job requirements and resource availability, including accessing the QC devices via a REST API for hybrid computations. The assigned QC device processes the jobs and returns the results and

---

[23] https://github.com/Simula-COMPLEX/quito.
[24] https://github.com/Simula-COMPLEX/qucat-tool.

resource usage information to SLURM, which then relays both the results and the metadata to the HPC system. Finally, the HPC system returns the computational results to the user, and reports the resource usage back to Puhuri for project accounting. This workflow ensures an efficient allocation of resources based on the job requirements, availability, and user entitlements, whilst managing the intricate process of coordinating between HPC and QC devices.

3.3   User Engagement

The NordIQuEst project kicked off officially in April 2022, and the infrastructure became operational in a testing phase in September 2022, in conjunction with the first NordIQuEst workshop. Since November 2022, a stable connection between LUMI and the VTT quantum computer Helmi has allowed for testing the HPC+QC infrastructure across the Nordics and Estonia. For example, the connection has been utilised in NordIQuEst organised events, such as the Quantum Autumn School 2023, co–organised with the EuroCC National Competence Centre Sweden (ENCCS)[25] and the Wallenberg Centre for Quantum Technologies. It is also used at the NeIC Conference 2024, Nordic e–Infrastructure Tomorrow, within the theme of these Proceedings.

The setup of the NordIQuEst platform has been shown to be stable and able to handle large loads, as tested during the Quantum Autumn School 2023, which is highlighted in Table 1. This shows the usage statistics over the short period of the event, where users were introduced to hybrid HPC+QC concepts, and were able to run their first hybrid HPC+QC jobs. The success of this event was aided by the Finnish Quantum–Computing Infrastructure (FiQCI),[26] which is one of the main synergistic efforts working together with the NordIQuEst project. The table displays the total number of users which used the platform over the course of the event, with the total number of jobs submitted and the number of shots used. The high number of shots, which is a single execution of an algorithm on the QPU, highlights the engagement of the users during this event.

Table 1. Statistics of HPC+QC jobs run for the Quantum Autumn School 2023, from October 25th 2023 to October 28th 2023.

Users	Total jobs submitted	Shots used
83	364	2533588

4   Challenges: Present and Future

The NordIQuEst infrastructure, while pioneering in its integration HPC and QC, has faced a multitude of diverse challenges, as expected for a conceptually completely new implementation. These challenges encompass technical, operational,

---

[25] https://enccs.se/.
[26] https://fiqci.fi/.

and conceptual domains, reflecting the complexity of merging these powerful but different computing paradigms. This chapter identifies the most prominent of the challenges faced, and offers some insights into conceiving potential solutions.

4.1   Technical Challenges

(a) **Quantum hardware variability**: The current landscape of quantum computing is full of varying emerging technologies, such as superconducting, ion trap and neutral atoms, each with its own unique strengths and limitations. This diversity has lead to significant variability among available quantum computers in terms of qubit count, error rates and operational capabilities. Addressing this challenge requires the development of an adaptable infrastructure, to match computational tasks to the most suitable quantum hardware. The NordIQuEst project has recognized the uncertainty surrounding quantum technology and aims to mitigate its impact by abstracting the hardware layer as much as possible. QC sites in the NordIQuEst project try and offer as much information about the quantum computing stack as possible to the HPC sites through a variety of cutting edge software solutions. Additionally, by offering a wide variety of simulators, with the classical resources to back it up, users can experiment and test out potentially different solutions.

(b) **Quantum hardware reliability**: Current NISQ quantum computers are delicate devices, requiring lots of calibration work and human maintenance to maintain uptime. This is in stark contrast to the HPC machines with which they are connected to, which provide users with very high up–times, availability and performance. Within the NordIQuEst project, it is the responsibility of QC sites to take care of maintenance and upkeep of the quantum computers, in addition to providing HPC sites with the availability information, such as whether the quantum computer is available for job submission and what is the current calibration status. Presently, for LUMI, this is solved in a straightforward manner with the automatic opening and closing of the SLURM job scheduler upon signal from Helmi.

(c) **Software instability**: The software enabling the use of quantum computation, be it simulator or real device, is in a very unstable state, reflecting the current state of quantum computing. Within the confines of a cross–border project it has proven to be a difficult challenge to maintain and update software packages, so that different software stacks are in sync and interoperable. The NordIQuEst project has learned that software installers are required to keep the software updated and to test new versions frequently. In addition, researchers are in close contact with maintainers in order to track potential problems and fix them promptly.

4.2   Operational Challenges

(a) **Quantum hardware availability**: Quantum computers are typically designed to be used by one user at a time, utilizing a first–in–first–out

(FIFO) queuing system at the level of the control electronics and lab equipment. Due to this, access to the quantum device is sequential, and notably, users may be accessing the device from different sources. Additionally, there needs to be a mechanism for users to reserve time slots for their quantum algorithms, to ensure exclusive access and the best use of their time. A sufficiently smart enough scheduler could be employed to manage the execution of quantum jobs which avoids stalling the quantum computer and efficiently plans future job execution. The scheduler should consider factors such as the current quality of the device, time constraints, and the demand for the quantum resources. An early solution, adopted within the FiQCI framework, was to set a daily time slot for HPC users and implement a hard limit to the number of job executions and time limit of jobs to avoid stalling. This required good communication between HPC and QC sites. The future hope is to have a dedicated quantum metascheduler, with a fair–share queuing system, that takes into account both the HPC system and the quantum computer. The implementation of an efficient co–scheduler is expected to be one of the most challenging tasks for HPC+QC infrastructures.

(b) **User quotas and project management**: The administrative tasks that are required for HPC style projects can be seen as a hindrance for access to experimental resources such as quantum computers. Estimating the amount of classical and quantum resources needed is difficult for researchers of hybrid quantum algorithms and complex processes for project management might prevent adoption from new researchers. User quotas impose restrictions on the amount of computational resources that individual researchers or projects can access. Experienced researchers with demanding computational requirements may find themselves constrained by these quotas, leading to delays in their work. To address these challenges organizations can organize workshops to deliver temporary collective access. While the practical experience among the user base is building up, it is important to provide a flexible resource application process, with short application processing times. Additionally, offering training through easy to follow tutorials and Jupyter notebooks can help new researchers familiarize themselves with quantum computing concepts and the tools needed to perform hybrid computation.

4.3   Conceptual Challenges

(a) **Hybrid applications assessment**: Properly evaluating quantum algorithms is a challenge due to the many heuristics involved. The nature of hybrid algorithms combining both classical and quantum components leads to additional complexity in terms of assessing how well they are performing, how well they scale, and the potential applications. The NordIQuEst project has found that many users prioritize the execution of quantum algorithms over in–depth analysis of benchmarking, testing the scalability and assessing the application. For many, it is important that it just executes. These challenges can be addressed by promoting evaluation studies that focus on the assessment of hybrid quantum applications. Providing software tools that facilitate the isolation of variables and automate the assessment of algorithms. An important component is to enable mitigation of the noise from NISQ–era quantum devices to enhance the reliability and reproducibility of results. Classical resources in HPC environments are in an abundance compared to their quantum counterparts. Hybrid applications should try to leverage the parallel computing power offered by HPC to enhance the assessment of hybrid algorithms.

(b) **User engagement** While the NordIQuEst infrastructure is still under development and not generally available to users, it is important to proactively consider user engagement. Our experience is that some clear barriers to user adoption exist. Some of these are related to insufficient knowledge about what an HPC+QC infrastructure can provide for a user, and others to the practicalities of getting access to it. As discussed in Sect. 3.2, the envisioned journey of a prospective user begins with applying for a project. This requires that the user articulates their research objectives in order to properly assess the computational quotas. Although necessary, this process might be hold back new researchers in adopting experimental technologies. As we have ascertained during NordIQuEst's events, facilitating this by providing access to the platform through a single project account and sharing its resources, significantly lowers this barrier. Especially for users who are not familiar with standard HPC procedures, providing a hands–on walk–through of the process is highly beneficial. Conversely, more experienced QC users might already have direct access to QC devices – such as those hosted by VTT or Chalmers – and might not resort to NordIQuEst's services unless they see an added value, e.g., in the form of hybrid computation or heavy pre– and post–processing. This added value thus needs to be clearly formulated and communicated to potential users. To engage users, it is also crucial to build up the infrastructure in constant rapport with the end–users. In practice, this means that the project implementation needs to be sufficiently flexible to allow for developing the user experience on the fly.

(c) **Education and training**: Both HPC and QC are complex domains that demand a level of expertise and experience to navigate efficiently. Integrating HPC with QC adds an additional layer of complexity, resulting in a steep learning curve for new users. Becoming proficient in both HPC and QC requires dedicated time and education in quantum computing, using a supercomputer, computer science, and additional domains. Bridging the gap between the HPC realm and the QC realm is crucial for enabling the adoption of hybrid infrastructure. This requires developing training programs and educational material tailored to equip researchers and potential users with the necessary skills and knowledge. Such training programs should include primers, tutorials, and plug–and–play demonstrations for both beginners and advanced users. In particular, the NordIQuEst project has seen the need for plug–and–play examples to introduce users and increase adoption. Advanced examples of hybrid algorithms are also needed for experienced users looking to optimize their workflows. Such examples should demonstrate the utilization of open–source tools and cutting–edge techniques to get the most out of the resources available.

5   Conclusions

The NordIQuEst collaboration has established a pioneering groundwork for integrating HPC and QC resources into a cohesive, hybrid computing infrastructure. This effort – the first of its kind on such a scale – was driven by the potential of offering researchers the opportunity to experiment with the novel technology of quantum computing, while accessing the capabilities of usual supercomputers.

Since its inception, the NordIQuEst platform has been predisposed with a diverse set of tools, enabling researchers to advance important scientific goals, such as evaluating variational hybrid algorithms, benchmarking current quantum computers, and implementing new hybrid applications. Nonetheless, the NordIQuEst project has faced, and successfully addressed, several technical challenges, including the need for an adaptable infrastructure that can accommodate the diverse nature of quantum hardware, and the software instability inherent in this rapidly evolving field. On the operational side, the project has managed to provide reliable access to quantum hardware, while leveraging existing systems and collaborative projects such as FiQCI (Research Council of Finland) and Puhuri (NeIC/Nordforsk) for coordinating user access and resource allocation. Furthermore, the project's commitment to cultivating a research community, has produced hands–on training events and educational materials, aimed at bridging the knowledge gap between classical HPC users and the QC practitioners.

In conclusion, while the NordIQuEst project has made significant strides in unifying HPC and QC systems, it is only an initial step towards a future where the synergistic potential of these technologies can be fully realized. Our journey underscores the importance of international collaboration in fostering innovation at the forefront of computing technologies, setting a precedent for future endeavors that will build on our foundational work. We end by highlighting the need for open–source development and open access material for enabling adoption and (re)utilization of the work performed.

Acknowledgments. We thank the NeIC/Nordforsk collaboration for providing funding and believing in the NordIQuEst project. We further appreciate the operational support offered by the Puhuri project. We acknowledge the invaluable contribution of every institution participating in the NordIQuEst collaboration.

References

1. Asadi, A., Dusko, A., Park, C.Y., et al.: Hybrid quantum programming with PennyLane Lightning on HPC platforms (2024)
2. Bayraktar, H., Charara, A., Clark, D., et al.: cuQuantum SDK: A High-Performance Library for Accelerating Quantum Science (2023)
3. Bergholm, V., Izaac, J., Schuld, M., et al.: Pennylane: automatic differentiation of hybrid quantum-classical computations (2022)

4. Cong, I., Choi, S., Lukin, M.D.: Quantum convolutional neural networks. Nat. Phys. **15**(12), 1273–1278 (2019). https://doi.org/10.1038/s41567-019-0648-8

5. Dallaire-Demers, P.L., Killoran, N.: Quantum generative adversarial networks. Phys. Rev. A **98**, 012324 (2018). https://doi.org/10.1103/PhysRevA.98.012324

6. Devoret, M.H., Schoelkopf, R.J.: Superconducting circuits for quantum information: an outlook. Science **339**(6124), 1169–1174 (2013). https://doi.org/10.1126/science.1231930, https://www.science.org/doi/abs/10.1126/science.1231930

7. DiVincenzo, D.P.: The physical implementation of quantum computation. Fortschr. Phys. **48**(9–11), 771–783 (2000). https://doi.org/https://doi.org/10.1002/1521-3978(200009)48:9/11<771::AID-PROP771>3.0.CO;2-E

8. Farhi, E., Goldstone, J., Gutmann, S.: A quantum approximate optimization algorithm (2014)

9. Filippov, S., Leahy, M., Rossi, M.A.C., García-Pérez, G.: Scalable tensor-network error mitigation for near-term quantum computing (2023)

10. Gray, J., Kourtis, S.: Hyper-optimized tensor network contraction. Quantum **5**, 410 (2021). https://doi.org/10.22331/q-2021-03-15-410

11. Johansson, M., Wendin, G.: The quest for a nordic quantum computing ecosystem. ERCIM News (128), 31–32 (2022). https://ercim-news.ercim.eu/en128/special/the-quest-for-a-nordic-quantum-computing-ecosystem

12. Johansson, M.P., Krishnasamy, E., Meyer, N., Piechurski, C.: Quantum Computing - A European Perspective (2021). https://doi.org/10.5281/zenodo.5547408

13. Jones, T., Brown, A., Bush, I., Benjamin, S.C.: Quest and high performance simulation of quantum computers. Sci. Rep. **9**(1), 10736 (2019). https://doi.org/10.1038/s41598-019-47174-9

14. Nakaji, K., Kristensen, L.B., Campos-Gonzalez-Angulo, J.A., et al.: The generative quantum eigensolver (GQE) and its application for ground state search (2024)

15. Navaux, P.O.A., Lorenzon, A.F., Serpa, M.d.S.: Challenges in high-performance computing. J. Brazil. Comput. Soc. **29**(1), 51–62 (2023). https://doi.org/10.5753/jbcs.2023.2219, https://sol.sbc.org.br/journals/index.php/jbcs/article/view/2219

16. Patra, S., Jahromi, S.S., Singh, S., Orus, R.: Efficient tensor network simulation of IBM's largest quantum processors (2023)

17. Peruzzo, A., McClean, J., Shadbolt, P., et al.: A variational eigenvalue solver on a photonic quantum processor. Nat. Commun. **5**(1), 4213 (2014). https://doi.org/10.1038/ncomms5213

18. Preskill, J.: Quantum computing in the NISQ era and beyond. Quantum **2**, 79 (2018). https://doi.org/10.22331/q-2018-08-06-79

19. Rebentrost, P., Mohseni, M., Lloyd, S.: Quantum support vector machine for big data classification. Phys. Rev. Lett. **113**, 130503 (2014). https://doi.org/10.1103/PhysRevLett.113.130503, https://link.aps.org/doi/10.1103/PhysRevLett.113.130503

20. Shor, P.W.: Fault-tolerant quantum computation (1997)

21. Smith, K.N., Viszlai, J., Seifert, L.M., et al.: Fast fingerprinting of cloud-based NISQ quantum computers. In: 2023 IEEE International Symposium on Hardware Oriented Security and Trust (HOST), pp. 1–12 (2023). https://doi.org/10.1109/HOST55118.2023.10133778

22. Tindall, J., Fishman, M., Stoudenmire, E.M., Sels, D.: Efficient tensor network simulation of IBM's eagle kicked ising experiment. PRX Quantum **5**, 010308 (2024). https://doi.org/10.1103/PRXQuantum.5.010308, https://link.aps.org/doi/10.1103/PRXQuantum.5.010308

23. Wendin, G.: Quantum information processing with superconducting circuits: a perspective (2023)

Federated Resource Allocation for HPC Services

Ilja Livenson[1] (ID), Ahti Saar[1] (ID), Marina Adomeit[2] (ID), Eivind Lysø[3],
Emma-Lisa Hansson[4], Jarno Laitinen[5](✉) (ID), and Anders Sjöström[4] (ID)

[1] University of Tartu, Ülikooli 18, 50090 Tartu, Estonia
[2] SUNET, Tulegatan 11, Stockholm, Sweden
[3] SIKT - Abels Gate 5, 7030 Trondheim, Norway
[4] LUNARC, Lund University, Box 118, 221 00 Lund, Sweden
anders.sjostrom@lunarc.lu.se
[5] CSC - IT Center for Science Ltd., P.O. Box 405, 02101 Espoo, Finland
jarno.laitinen@csc.fi

Abstract. This paper presents an implementation of a federated resource allocation system (Puhuri), which offers a full set of functions from the application phase for computing infrastructure resources to consumption reporting. The implementation is based on the Puhuri project of Nordic e-Infrastructure Collaboration (NeIC) [1]. Shared research infrastructures, particularly high-performance computers, necessitate a robust mechanism for allocating resources in accordance with predefined policies. An allocation serves as authorisation to utilise a system in an agreed-upon manner. The impact of Puhuri on shared research infrastructures is to enable multiple resource allocation teams to allocate resources across different domains and organisations. Puhuri has integrated the LUMI EuroHPC supercomputer [2] and the biodiversity digital twin workflow management framework Lexis. Additionally, Puhuri has also been collecting requirements and doing technical integrations in a wider scope. Users accessing the system must be unequivocally identified. This process, known as authentication, ensures that the right individuals utilise the allocated resources to which they are authorised by the resource allocation team. GÉANT has established the MyAccessID service [3] for federated authentication, which Puhuri also utilises. One of the challenges is related to the user identity vetting for the users whose identity provider is not signalling that they are verifying users' identities in the required manner, which requires them to use commercial third-party services for automatic identity vetting based on user's identity cards.

Keywords: Accounting · Authentication · Authorisation · Federation · Governance · HPC · Identity vetting · Resource allocation

1 Introduction to Resource Allocation

1.1 Resource Allocation in High-Performance Computing

Large-scale research infrastructures represent significant investments. It is in the interest of the stakeholders in the research infrastructure e.g. funders, resource allocators, resource operators, and users that the resource is utilised as efficiently as possible and that

© The Author(s) 2025
A. Azab and T. Malkiewicz (Eds.): NeIC 2024, CCIS 2398, pp. 128–141, 2025.
https://doi.org/10.1007/978-3-031-86240-3_9

the access to the resources follows the policies of the funders, allocators, and resource operators while being perceived as fair by the users.

A resource in this context should be understood as a service, which requires permissions to access and has consumption limits for the usage. The right to access a resource is defined through an allocation of a Project on the resource with a start date and an end date and to the extent of a number of consumable units. In the high-performance computing (HPC) community, a good utilisation rate and fairness have been achieved through allocations of consumable units on the resources e.g. CPU core-hours, which are then consumed on the resource via a queueing system e.g. Slurm, PBS, LSF etc.

The Puhuri federated resource allocation and reporting system provides the stakeholders mentioned above with a one-stop shop for performing all of the tasks related to resource allocation and utilisation, combined with a reporting interface where the stakeholders can keep track of the allocated and consumed resources. The federated aspect is in integrating multiple identity providers on the authentication layer via the MyAccessID identity proxy and also on the allocation layer where multiple resources can be allocated.

1.2 Actors in Resource Allocation

In the resource allocation process there are different roles:

1. **Funders**: who might set policies on who can use resources and how much. The calls are defined based on these principles. Funders usually want to see the impact of the funding, and therefore it is important to collect different statistics about the project proposals, resource allocation requests and resource usage.
2. **Resource allocation teams**: review the applications based on the policies and regulations of the funders and resource providers. The target is the fair allocation of resources to the users based on the information and metrics supplied by the applicants and the resource providers. The resource allocation team may also be tasked with aggregating and producing reporting for the funders.
3. **Resource providers**: who provide the actual resources and support the users in various ways. Resources could vary from storage and web portals to large-scale high-performance computers or other infrastructure services. Resource providers provide information about the resource consumption and utilisation for the resource allocators and funders. Collaboration is needed to ensure security and adherence to regulations limiting access to the system.
4. **Users**: who through the Principal investigator apply for resources for the team in the case that there is a project-based allocation. All users must adhere to the Terms of Use that the service sets up, and use the resource exclusively for the purpose expressed in the application.

Puhuri sets up usage contracts with the resource allocation organisation and the resource providers. To cover the operational costs of the Puhuri system there is a fee. Puhuri is built on open-source components. In particular, the Puhuri Portal can be both self-hosted or requested as a managed service, operated by the University of Tartu.

Puhuri assists the users and organisations in an agreed manner via the Puhuri helpdesk and web-based documentation [4]. However, the primary responsibility of the Puhuri user support is to support the resource providers and allocators. Thus, Puhuri may in practice be invisible to the end users.

1.3 The Resource Allocation Units

The resource allocation unit can be system-specific. The most basic are core-hours i.e. the right to use 1 CPU core for one hour. This is the allocation model for the LUMI supercomputer [2]. Additionally, LUMI has the concept of storage hours corresponding to the right to store a set of data of the size of 1 GB for one hour. A different model could be to define storage quotas for the project lifetime.

Allocation units may also take the cost or scarcity of a resource into account, this is often done by setting different consumption rates. For example, the consumption rate of a GPU may be larger than for a CPU. Some resource providers use a common resource allocation unit for heterogeneous hardware. For example, "billing units" are used by CSC - IT Center for Science Ltd. [5] different CSC's HPC, cloud and object storage systems use the same billing unit. In that manner user needs to apply for only one kind of allocations unit to be consumed for various services with different amount of billing units per hour. Similarly "service units" are used by the University of Maryland [6] and the "allocation unit" by the National Renewable Energy Laboratory [7]. ACCESS [8] is an advanced computing and data resource program supported by the U.S. National Science Foundation (NSF), and it uses "grants" and "credits", which can be exchanged for different resources based on the particular conversion rate.

There is also the possibility of using a pure bartering scheme where resource allocation units are bartered between resource providers thereby providing access for users of one resource to resources of another. As shown in the NeIC project Dellingr [9], doing this across borders i.e. international resource exchange, means that the bartering parties need to consider value added taxes [10].

1.4 The Project Concept

The resource allocation is typically done for a research project consisting of members. The project's Principal Investigator (PI) or co-PI, apply for a project and manage project members by inviting or removing them to the project via a portal or other support channels. The project membership may also allow access to the project's shared data as well. Therefore, trust in the applicant's identity must be established to avoid mistakes in the invitation.

The user of a system usually needs to register to the allocated service to get a personal account. The Users may then also need to accept an Acceptable Use Policy or Terms of Use document as an authenticated user. This may all be part of the granting process.

1.5 The Allocation Types

The resource allocation policies of the respective resource-allocating teams and funders may also define different allocation categories e.g. development, benchmarking, regular, or extreme scale access as defined by the EuroHPC JU [11], to which the project can apply for access. The terms of the allocation policies may influence who can apply, the duration of the allocation and how many resources can be applied [12]. There may also be prioritised queues and related allocations for urgent or important projects.

1.6 Central or Distributed Decision-Making Models

The following list uses a portal as a channel to make the resource allocation decisions to the service providers:

1. Local portal for a service provider. Example. my.csc.fi for CSC - IT Center for Science Ltd. to allocate resources for the projects created on the portal.
2. Central portal for multiple service providers: The EuroHPC review portal is an example, although it is not currently programmatically integrated with the resource providers. Puhuri Portal can be used for multiple resource providers' allocation types.
3. Federated portals: Many portals can be used to grant permissions for one or more resource providers.

For example, the LUMI EuroHPC supercomputer [2] is owned by the European High Performance Computing (EuroHPC) Joint Undertaking [11] and numerous countries. Therefore, the resource allocations are coordinated along the agreed principles and based on the funding of the owner organisations.

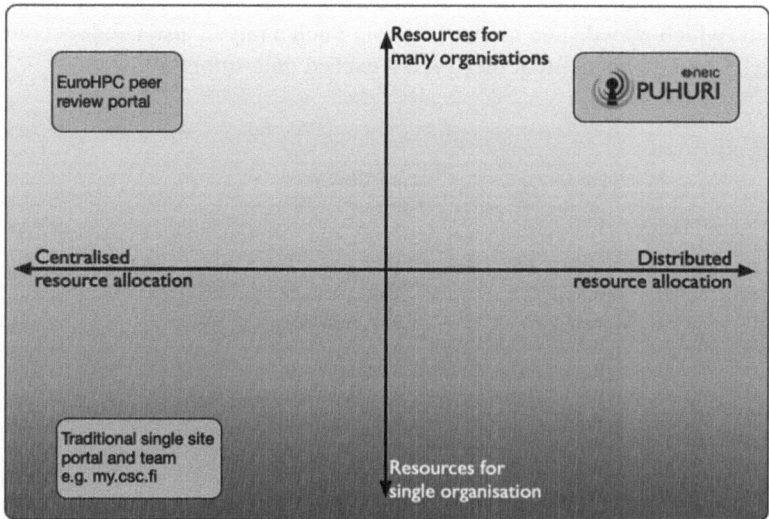

Fig. 1. The resource allocation decisions can be made by single or multiple resource allocation organisations, which may use one or more resource allocation portals. There can be one or more resources as a target system.

As illustrated in Fig. 1, Puhuri can integrate multiple resource allocation portals of one or more resource allocation teams to allocate resources for one or more resources for different allocation types.

The Fenix consortium, consisting of six major HPC centers in Europe, together with a software company created a federated resource allocation portal called FURMS [13]. The main difference between Puhuri and FURMS is that FURMS is a centralised portal whereas Puhuri is distributed. The ACCESS [8] portal for advanced computing and data resource program supported by the U.S. National Science Foundation (NSF) has a public catalogue of its services as well as a wizard to choose from based on the needs. The ACCESS Allocations Review Committee (AARC) is set to review the applications.

Before the allocation request is decided upon, there may be a more or less complex review process applied to the request. The more resources are needed, the more carefully the request should be analysed so that the planned consumption makes sense and is fair toward other users as there often is competition for somewhat scarce resources. How the call is implemented and integrated into the resource allocation process varies. In the simplest case, there is no extra administrative work, and the decision to grant resources would be made without any additional steps or even automatically. To meet the needs of resource allocators, Puhuri is implementing an application review support function. It is also possible to attach existing review portals directly into Puhuri Core via an API, as has been done with the Swedish SUPR portal [14].

1.7   Hierarchical Allocation Models

There may also be umbrella models, which have a share to be allocated for the projects (see Fig. 2). The umbrella can be an organisation, national, or research infrastructure community, which may decide their allocation. Such a model also needs accounting on the umbrella level to ensure that they do not exceed their quotas.

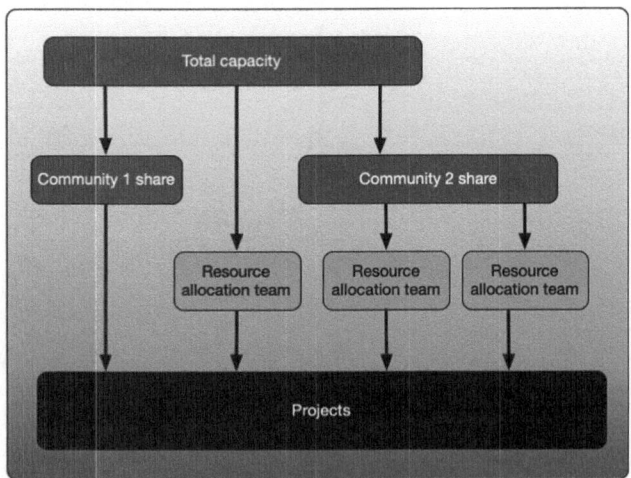

Fig. 2. The resource allocations can be distributed based on different models.

2 Puhuri

2.1 The Puhuri Project

The Puhuri project was launched in mid-2020 by the Nordic e-Infrastructure Collaboration (NeIC). The key driver for initiating the Puhuri project was the need for an authentication and allocation infrastructure for the EuroHPC LUMI supercomputer. LUMI is hosted in Finland, and the system is co-owned by a consortium of eleven countries and the EuroHPC JU. Each owner needed to be able to manage allocations individually for communities of users coming from different countries and sectors, so it was clear that the partners needed a common way to authenticate, authorise and identify researchers who accessed the system. It was decided that a suitable AAI would be developed. The Puhuri project started a collaboration with GÉANT eduTEAMS [15] resulting in MyAccessID, where also the FENIX HPC infrastructure was integrated.

The Puhuri project builds upon the experience from an earlier NeIC project, known as Dellingr [9], which investigated the feasibility of sharing and exchanging HPC resources (such as computing, storage and support services) between countries and across borders.

2.2 Technical Architecture

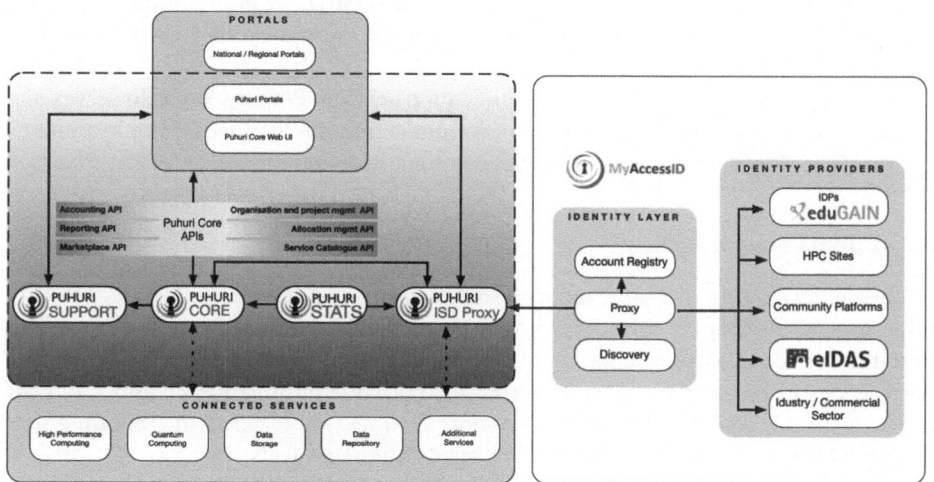

Fig. 3. The Puhuri technical architecture.

The Fig. 3 shows the components of Puhuri and MyAccessId and the connected services.

2.3 Puhuri Portal

The Puhuri Portal is a web-based portal for resource allocators and end-users integrated with Puhuri Core through APIs. The portal allows for multi-tenant management of projects, teams, and resource allocations. The Puhuri Portal is also integrated with Puhuri Support and can be customised and white-labelled to match the branding requirements of the resource allocator.

The Puhuri project is continuously working to improve the user experience and expand the functionality of the Puhuri Portal, providing resource allocators and users with a modern and convenient tool.

One of the current ongoing extensions is the introduction of allocation call management and application reviews. It is about the resource allocators' policy if they open calls for the applications or if the resources can be applied at any time. This is communicated on the web portal or other communication means.

The Puhuri Portal is available both as self-hosted software and as a managed service, which the University of Tartu operates.

In addition to using the Puhuri Portal, resource allocators have the option to integrate their existing portals with Puhuri Core. This has been done with the Swedish SUPR portal [14] and the LUMI projects allocated in a Finnish portal are also pushed to Puhuri Core.

2.4 Puhuri Core

Puhuri Core is a resource allocation service that connects resource allocation portals and service providers. One example of such integration is with the LUMI EuroHPC supercomputer, which is seamlessly integrated with Puhuri Core in an automated manner. This allows for the efficient management of LUMI resources by the LUMI consortium members in a fully digital way.

In addition to its role as a resource allocation service, Puhuri Core also serves as a channel for the propagation of information from services to end-users. For example, it can provide information on the status of allocations, accounting data, or local usernames.

2.5 Puhuri Statistics Portal

Puhuri is developing a central reporting platform for service providers and resource allocators, where they are able to monitor resource allocation and usage information. This portal is located at https://puhuri-stats.neic.no. Access is granted based on the project permissions in Puhuri Core. Puhuri portals also provide information on allocation and usage. Still, this central solution offers a more flexible way to select the reporting time range and better information presentation across all the projects.

For getting the reporting information from the service provider, Puhuri runs a separate microservice, which pulls usage data from the service provider's accounting database. Then, this usage data is exposed to the portal users as well as to Puhuri statistics environment users. Puhuri also supports the option where the service provider can push usage data to Puhuri Core by using a dedicated API endpoint.

2.6 Puhuri Technical Documentation

The website https://puhuri.neic.no [4] provides information for both resource allocators and service providers. Resource allocators can access a user guide that includes screenshots of the Puhuri Portal, as well as integration instructions for those who wish to integrate their existing national portal with Puhuri Core. On the other hand, service providers can find information on how to integrate their HPC systems with Puhuri Core.

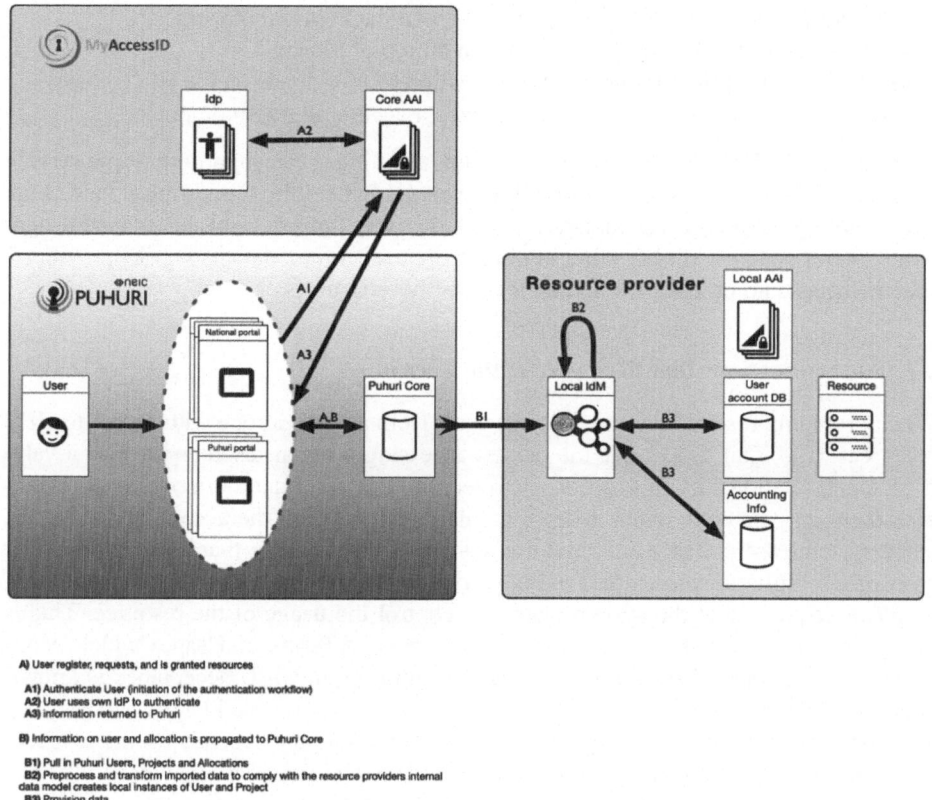

A) User register, requests, and is granted resources

A1) Authenticate User (initiation of the authentication workflow)
A2) User uses own IdP to authenticate
A3) Information returned to Puhuri

B) Information on user and allocation is propagated to Puhuri Core

B1) Pull in Puhuri Users, Projects and Allocations
B2) Preprocess and transform imported data to comply with the resource providers internal data model creates local instances of User and Project
B3) Provision data

Fig. 4. Integration of the portals and resource providers to Puhuri.

The Fig. 4 shows how the user authenticates to a resource allocation portal with the identity provider the user selects in MyAccessID Idp Proxy (A1-A2). If the user has not yet registered to MyAccessID, the user is forwarded there to register in MyAccessID. User information is passed to Puhuri Core via Portal (A3 and A, B).

The project may already exist, or it will be created. The PI might invite members to the project. Once the resource allocation team accepts the request, it is passed to Puhuri Core (A, B), where the resource providers are integrated. Those will receive the projects with members, possibly with the ssh keys, and allocations (B1). The resource provider implements its user administration and accounting system in its own way, including local user account generation (B2). Services such as SSH and batch scheduling systems are integrated there. Puhuri has also implemented an API to pass the local user account and accounting information back to the portal via Puhuri Core (B3). Puhuri Core provides APIs to integrate any kind of system by the HPC resource provider [4].

Puhuri Core provides ready to use open source integrations with popular HPC and Cloud software stacks including:

- Slurm—one of the most popular job schedulers;
- OpenStack—open-source IaaS solution for running on-premise cloud;
- Rancher Kubernetes to run K8s on top of OpenStack virtual machines;
- CSCS FireCrest—API for job submission on top of Slurm.
- HEAppE - REST API for controlled execution of workloads on the HPC cluster
- Open OnDemand - easy-to-use web portal for accessing supercomputers

Puhuri Core is designed to be multi-tenant, meaning it can support multiple service providers and services simultaneously. It is also highly flexible, supporting a wide range of accounting models. These models can include fixed, upfront, limit-based, and usage-based models. This flexibility allows Puhuri Core to accommodate its service providers' specific needs and preferences and their respective end-users.

2.7   Authentication: Identifying Who the User is

When providing researchers and industry users from across Europe with access to HPC resources, it is vital to identify the person accessing a given resource to have a valid contractual relationship. In addition to protecting data and unauthorised access in general, there may be contractual or legal regulations to control the access. For example, some systems are classified as dual-use systems [16] meaning that individuals under embargo or affiliations in embargoed countries are restricted from the use of the product. This requires that the system operators control the usage of the resource. This is partly achieved through the acceptable usage policy (Terms of Usage) which needs to be accepted and adhered to by the users. Central to any ToU acceptance is knowledge of the identity of the user and the level of assurance on the identity of said user. The required level of assurance (LoA) of the user's identity depends on the resource provider's requirements, meaning that different resource providers may have different requirements. Ideally, all resources should have the same requirements on LoA, but this is not always the case. The Puhuri system and MyAccessID have been prepared so that the LoA can be provided if requested.

Within individual countries, there are examples of national R&E identity federations with thorough home organisation coverage e.g. WAYF, HAKA, FEIDE, and SWAMID in the Nordics. However, on the European level as a whole, not every country has an all-encompassing identity federation with good coverage.

The Puhuri project has been working with GÉANT and the FENIX infrastructure to develop the MyAccessID, an Identity and Access Management Service, which provides a common identity layer for Infrastructure Service Domains (ISDs) - in other words, MyAccessID enables researchers to use their home organisation IdP or similar, to identify themselves with the appropriate level of assurance so they can easily access HPC resources located elsewhere. MyAccessID is available for all users whose identity provider is published in the inter-federation eduGAIN [17].

2.8   Accessing Provided Resources

In order for users to access the resources, they need to follow through the process of registering at MyAccessID which typically happens during the resource allocation process. During that process, the users belonging to the accepted projects would typically be sent an email asking to accept the invitation and login through the relevant allocation portal. After the user is part of an accepted project proposal, the information about the project, the allocation and the users would be recorded in the Puhuri Core, and then forwarded to LUMI's local Identity Management system which controls the access rights to LUMI.

Access to LUMI resources is done with the use of the SSH keys. Users need to upload their keys to their MyAccessID profile. The SSH key is then synced with the user's information stored in the Puhuri Core and further with the LUMI local Identity Management system.

Beyond personalised accounts, Puhuri Core supports the creation of 'robot accounts' that represent a subset of users in a project or can have their own dedicated credentials. Creation and management of such accounts is done by Service Providers, requesting and granting such accounts is currently happening outside the Puhuri system.

Robot accounts are useful when a process cannot be linked directly to a specific user, for example, for CI/CD pipelines executed automatically. Another use case where robot accounts are useful is services that do not want to process any personal data and hence operate only with robot accounts with accounting data being linked to Puhuri resources.

3   Discussions

3.1   Impact of Puhuri

Puhuri has enabled a unique model for the LUMI EuroHPC resource where allocation by many research allocation organisations is possible. Puhuri's main key feature is the ability to connect multiple resource allocators to multiple resource providers. This provides an opportunity to save on the resource allocation administration expenses and project membership management portals and allows the federated login.

As the Fig. 5 and Fig. 6 show, the Puhuri uptake has been successful.

By using federated authentication, the users do not need to remember yet another password and the user identity vetting information source is delegated to the identity provider.

3.2   BioDT Use Case

The Biodiversity Digital Twin project (BioDT) [18] provides advanced models for simulation and prediction capabilities through practical use cases addressing critical issues related to global biodiversity dynamics.

The project has implemented a multi-tenant platform for digital twins built to support diverse use cases, in particular, support for dynamic allocation of resources in the presence of multiple resource allocators and multiple instances of digital twins. BioDT users

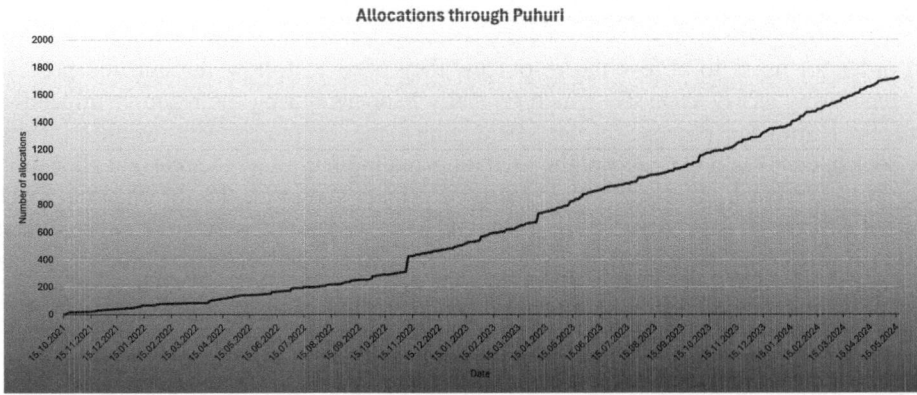

Fig. 5. Puhuri has been used for thousands of allocations by 2024.

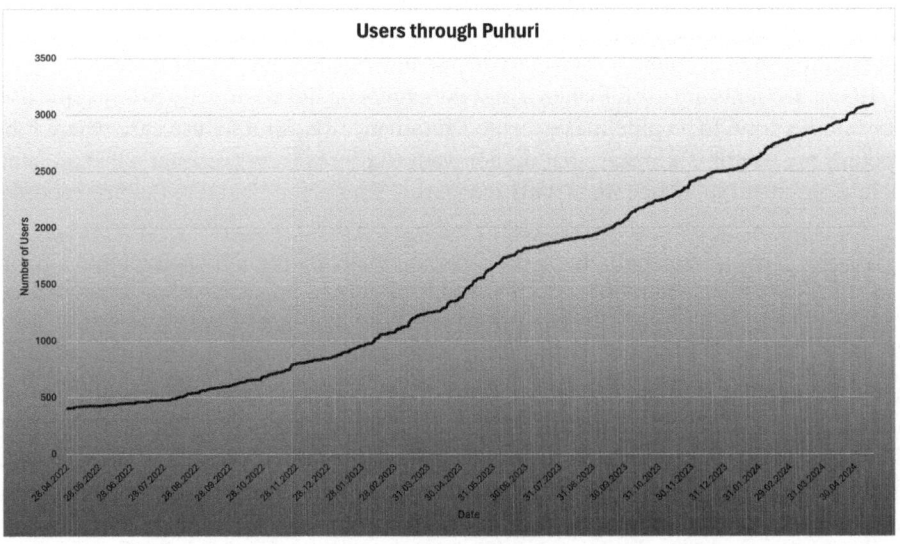

Fig. 6. Puhuri has been used by thousands of users by 2024.

can execute preconfigured workflows leveraging HPC infrastructure without accessing the HPC infrastructure.

The BioDT project relies on MyaccessID to provide global authentication and Puhuri to provide resource allocation services. Integration with Puhuri is intended to be as non-intrusive as possible and aimed to complement the existing HPC site's allocation policies and procedures while providing a common abstraction layer that higher-order orchestrators such as the LEXIS Platform [19] can rely upon.

Service accounts enable trustworthy indirect access to execute jobs for the HPC systems via the workflow orchestrator. Such functionality is important for easier integration of HEAppE middleware [20] used by LEXIS for running federated workflows combining multiple HPC and/or Cloud resources.

One of the key goals is to achieve good utilisation of the system while still being fair to the users. Therefore intelligent schedulers are built on the batch scheduling system or meta scheduler to send jobs for different systems, where the user has permission to. Puhuri does not contain such kind of scheduling logic, but a middleware layer such as Lexis to distribute the jobs and data.

3.3   AAI - Level of Assurance (LoA)

For a user to successfully access services connected to MyAccessID, IdPs are asked to release a minimal set of attributes about the user, i.e.; unique identifier, name, email, affiliation and home organisation [3]. Also, the reliability of the user's identity is signalled by the IdP through the Level of Assurance attribute. It may be that not all IdPs release affiliation and home organisation as that information might also be missing in eIDAS [21].

It depends on the service and how important it is to have high reliability against unauthorised use of resources and/or access to data. Level of assurance requirements are defined for two aspects of the identity assurance, using the REFEDS Assurance Framework [22], which is a global standard recognised in the R&E federation space:

- Identifier uniqueness to ensure unambiguous identification of users;
- Identity proofing and credential issuance, renewal, and replacement are needed to ensure that identity trustworthiness represents the right natural person.

The level of assurance is expressed at the time of user authentication by the IdP ending values defined in the RAF suite inside the eduPersonAssurance attribute.

For Puhuri and LUMI, it is important to identify users uniquely and have confidence in their trustworthiness. The requirements for access to Puhuri and LUMI is to express the level of assurance in the eduPersonAssurance attribute with the following values defined in RAF [22]:

- https://refeds.org/assurance/ID/UNIQUE   or   https://refeds.org/assurance/ID/eppn-unique-no-reassign for identifier uniqueness, and
- https://refeds.org/assurance/IAP/medium   or   https://refeds.org/assurance/IAP/high for identity assurance.

Enforcing such a requirement for IdPs to release a Level of Assurance information has proven to be challenging due to the fact that many of the Identity Providers currently do not release this information or the assurance level is too low. To this day, the Puhuri use case was the first and only international use case for the formulated Level of Assurance. In collaboration with MyAccessID, efforts towards the European IdPs are being made to urge them to meet the requirements for LoA, while alternative solutions are being explored in parallel. Such alternative solutions would enable users to use their organisational identity while they go through a process to elevate the LoA of that identity. One way of elevating the LoA would be to log in using their national eID and link it to their organisational identity. However, at this moment, national eID through the -framework covers only 19 European countries, which creates the need to use solutions for identifying users through passports, which are currently possible through integration with commercial providers. These solutions are being explored as well to be integrated into MyAccessID. Mobile application based EU Digital Identity (EUDI) Wallet [23] is coming into wider usage later.

3.4 Sustainability and Governance

One of the main focuses of the second phase of the Puhuri project was governance and sustainability issues. These are crucial factors in determining the project's long-term success and for eventually handing over the project to a final receiver. The project selected a legal entity to make contracts on behalf of the project during the project's lifetime. Also the IPR issues were reviewed with a legal expert. For the project's long-term success, strategies for the main service areas of Puhuri have been categorised containing strategies and goals for governance, technical operations and sales. A roadmap has been constructed for the continuation of the project, stating responsible parties and timeframes.

4 Conclusions

Puhuri has been successfully used for distributed resource allocation for different service providers, and the end-users have been either using their national portal or Puhuri Portal instance to manage project memberships and do the resource applications. At the time of writing, over 3000 users and 1800 project allocations have gone via Puhuri. The users have been using their home organisation's identity provider for authentication, and they have registered SSH public keys in the MyAccessID service to be distributed to resource providers. The Puhuri statistics portal has been developed based on the resource allocation reporting needs.

We expect GÉANT to provide centralised improved solutions for identity vetting and actively encourage in eduGAIN federation and AARC communities identity providers to provide a level of assurance attributes. The final solution for the user self-registration is still open. Puhuri aims to get several new customers and to ensure sustained operations without project funding from NeIC. Puhuri's near-term work aims to pilot peer review tools with different user organisations. We aim to provide means to accept service-specific terms of use in the Puhuri Portal or national portals.

Acknowledgments. The Puhuri project is funded by Nordic e-Infrastructure Collaboration (NeIC) and project partner organisations.

References

1. Puhuri. Nordic e-Infrastructure Collaboration project. https://neic.no/puhuri/. Accessed 20 Mar 2024
2. LUMI EuroHPC Supercomputer. https://www.lumi-supercomputer.eu/. Accessed 22 Mar 2024

3. MyAccessID web page. GÉANT association. https://wiki.geant.org/display/MyAccessID/MyAccessID+Home. Accessed 22 Mar 2024

4. Integration guide for service provides. Puhuri technical documentation. https://puhuri.neic.no/service-providers/. Accessed 22 Mar 2024

5. How CSC service use is measured and monitored. CSC - IT Center for Science Ltd. https://research.csc.fi/resources. Accessed 22 Mar 2024

6. Glossary for High Performance Computing Terminology. University of Maryland. https://hpcc.umd.edu/hpcc/help/glossary.html#su. Accessed 22 Mar 2024

7. System Resource Allocation Units. The National Renewable Energy Laboratory of the U.S. Department of Energy. https://www.nrel.gov/hpc/system-resource-allocation-unit.html. Accessed 18 Mar 2024

8. ACCESS Allocation website. https://allocations.access-ci.org/our-team. Accessed 15 Mar 2024

9. Dellingr. NeIC project home page. https://neic.no/affiliate-dellingr/. Accessed 15 Mar 2024

10. Resource exchange implementation and agreement. Dellingr project deliverable 5 (2020). https://wiki.neic.no/w/ext/img_auth.php/8/8a/NeIC_Dellingr_DO5_final.pdf

11. Access Policy of the EuroHPC Joint Undertaking Supercomputers

12. https://eurohpc-ju.europa.eu/system/files/2022-03/Decision%2018.2021%20-%20Access%20policy.pdf. Accessed 11 Mar 2024

13. FENIX User and Resource Management Service (FURMS). https://fenix-ri.eu/infrastructure/services/furms. Accessed 25 Mar 2024

14. GÉANT eduTEAMS. https://eduteams.org/. Accessed 14 Mar 2024

15. SUPR Portal. The National Academic Infrastructure for Supercomputing in Sweden. https://supr.naiss.se/round/. Accessed 25 Mar 2024

16. Dual use export-regulation. https://eur-lex.europa.eu/EN/legal-content/summary/dual-use-export-controls.html. Accessed 20 Mar 2024

17. eduGAIN interfederation service. https://edugain.org/. Accessed 25 Mar 2024

18. Biodiversity Digital Twin (BioDT). https://biodt.eu/. Accessed 25 Mar 2024

19. The LEXIS platform web page. https://lexis-project.eu/web/. Accessed 25 Mar 2024

20. HEAppE framework. https://heappe.eu/. Accessed 25 Mar 2024

21. eIDAS home page. European Commission (2023). https://digital-strategy.ec.europa.eu/en/policies/discover-eidas. Accessed 20 Mar 2024

22. REFEDS assurance suite (2023). https://wiki.refeds.org/display/ASS. Accessed 20 Mar 2024

23. The EU Digital identity. EU Commission. https://commission.europa.eu/strategy-and-policy/priorities-2019-2024/europe-fit-digital-age/european-digital-identity_en. Accessed 20 Mar 2024

From Isolation to Integration: A Decade of Container Technology in Slovenian HPC

Dejan Lesjak[1,3]([✉]), Alja Prah[1,2,3], and Barbara Krašovec[1,3]

[1] Jožef Stefan Institute, Jamova 39, Ljubljana, Slovenia
dejan.lesjak@ijs.si
[2] National Institute of Chemistry, Hajdrihova 19, Ljubljana, Slovenia
[3] Slovenian National Supercomputing Network SLING, Ljubljana, Slovenia

Abstract. This article examines the administrative strategies for managing software required by users and scientific communities within Slovenian Grid and HPC sites. It reviews the historical approaches to software enablement and transitions to a discussion on the adoption of software modules and containers. These methods are currently favored for their efficiency in ensuring the portability, performance, and security of scientific software. Additionally, the paper outlines the existing state of software deployment, utilising modules and container images across Slovenian sites, and proposes future plans aimed at enhancing security and deployment practices.

Keywords: Containers · Software modules · Portability · Performance · Security · Deployment

1 Introduction

Scientific software is complex and difficult to install on HPC systems. The installations are often non-standard and can be time-consuming due to the intricate web of dependencies required. These dependencies are frequently version-specific, necessitating precise library versions, which can be challenging, especially if the documentation is poor, outdated, or non-existent. Furthermore, the codebase itself may be poorly written, leading to inefficiencies and difficulties in parallelisation. Legacy code is especially difficult to maintain, optimise and parallelise, it is often not straightforward to install on modern operating systems.

Several key aspects became evident during the history of sites comprising Slovenian National Supercomputing Network (SLING)[1]. Given the disparity of hardware (CPU architecture in particular) and operating systems installed on physical worker nodes of clusters, the most pressing one at the beginning was portability. Going forward, to exploit the equipment fully and efficiently, there was concern about performance and the question of how much overhead, if any, the solutions for portability add. As the popularity of clusters among users

---

[1] https://www.sling.si.

© The Author(s) 2025
A. Azab and T. Malkiewicz (Eds.): NeIC 2024, CCIS 2398, pp. 142–151, 2025.
https://doi.org/10.1007/978-3-031-86240-3_10

increased, so too did concerns about security, which influenced the choice of the container solution. With the growth of computational power among HPC sites, they became a valuable resource for industry users, who have another set of requirements and computational needs.

This article presents the methodologies employed by SLING to address the aforementioned requirements and the prospective solutions to enhance the user experience on our HPC clusters.

2 History of Containers and Software on Slovenian Clusters

Due to collaboration on CERN's Large Hadron Collider (LHC)[2] experiment and consequently on The Worldwide LHC Computing Grid (WLCG)[3], Jožef Stefan Institute set up a cluster SiGNET[4] in 2003. It consisted of 51 nodes with two processors each and could be replaced by a single node today. It had connectivity of 2.5 Gb/s and utilised shared filesystems like AFS (Andrew File System), NFS (Network File System) and GFS (Global File System). At the time it was one of the first built with AMD 64-bit processors, which influenced the choice of operating system: it was decided that SuSE was the best choice as its support for the hardware architecture was most mature. Due to higher customisability it was later replaced by Gentoo. As some services for interfacing with WLCG needed to run on different operating system and sometimes completely in 32-bit mode, it was necessary to provide some sort of emulation and/or virtualisation. The scientific software running on worker nodes as well expected a specific operating system environment and architecture.

While the main services could run in virtual machine environments, a more lightweight environment was required for compute jobs within the cluster in order to minimise the overhead and enable the necessary environment for specific workloads. The story of container-like environments thus began with chroot and several wrapper and helper scripts, which were collectively referred to as Grid-COSI (Grid enabled Chroot Operating System Image). This enabled WLCG jobs from the LCG ATLAS[5] experiment to run in a 32-bit Scientific Linux environment, while natural language processing jobs ran in a 32-bit Debian setup. However, the scripts soon became unwieldy to maintain, prompting a migration to uchroot and later schroot[6], which is used by Debian for package building.

To address the challenges of building and maintaining scientific software for such environments, tools like EasyBuild [1,6] and Spack [3] have been developed. EasyBuild is a software build and installation framework that automates the process of managing software on HPC systems, ensuring consistency and repro-ducibility across different environments. The deployment of complex software

---

[2] https://home.cern/science/accelerators/large-hadron-collider.
[3] https://wlcg.web.cern.ch/.
[4] https://www-f9.ijs.si/egee/signet.html.
[5] https://atlas.cern/.
[6] https://wiki.debian.org/Schroot.

stacks is simplified by the handling of dependencies and configuration specifics. Similarly, Spack is a flexible package manager designed for HPC environments, enabling users to easily install and manage multiple versions of software and their dependencies. Spack's versioning and customisation capabilities allow researchers to tailor software installations to their specific needs while maintaining compatibility with the underlying hardware. These tools significantly reduce the overhead associated with manual software management, allowing scientists and developers to focus more on research and less on system administration.

Tools for custom environments continued to evolve in direction to provide them in more isolated way in form of containers. While popularity of Docker containers grew, dependency on root owned daemon was not best fit for jobs on cluster [5]. Singularity[7] project began as open source in 2015 [8]. We soon began exploring ways to improve security and make maintenance easier. Singularity uses definition files to create containers that run jobs. This makes it easier to upgrade and rebuild the environment. Definition files also help us document the environment and describe how it is set up. Together with the ability to convert a Docker container into a Singularity image, users can provide their own recipes or containers to run their jobs in, giving them more flexibility and control of the execution environment. On the other hand, isolation of different users' tasks improved the security of clusters.

3   Optimization of Software Environments

With rising awareness of available compute power, other groups were increasingly interested in running highly parallel workloads on existing centres. This prompted sites such as Arnes[8] and NSC[9] to focus on improving the MPI environment by deploying HPC interconnects from 2011 and 2015 respectively.

Performance is the cornerstone of HPC as it enables the processing of large-scale computations and complex problem-solving. Vendors and HPC/HTC centres tend to prioritise achieving the best performance, even if this means neglecting other aspects of running workloads on the system such as software portability or security.

In the past, there was a perception that containers were less efficient and caused performance overhead [11]. This was likely due to a misunderstanding of the underlying technology, which was being compared with virtual machines, where a hypervisor manages the virtual instances, consuming some of the system resources. However, subsequent measurements and benchmarking of workloads running on bare metal machines or in containers demonstrated that this consideration can be disregarded [7]. Containers are like any other application, a process on the system. Containers are very efficient, which is particularly notable

---

[7] https://sylabs.io/.

[8] https://www.arnes.si.

[9] https://www.ijs.si/ijsw/nsc.

in areas such as CPU and memory usage, where the overhead of running a container is negligible [9]. Also the choice of the container solution has minimal impact on the application performance [10].

For the EuroHPC Vega system[10], we have prepared a template container definition, consisting of MLNX_OFED[11], which can be used as a template for building applications on top of it. Running OSU benchmark[12] in the container and on the bare metal compute node showed that we can achieve almost identical results, as shown on Fig. 1.

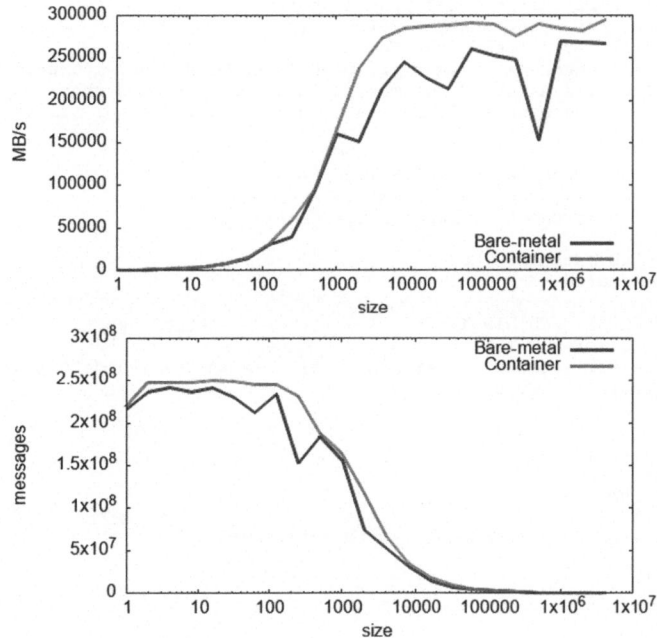

Fig. 1. OSU benchmark pt2pt/osu_mbw_mr: comparison of running the benchmark on a bare-metal server and in a container

We have used a hybrid MPI mode of running the container, which means that the Mellanox OFED was installed in the container image, the version in the container was identical as the version on the compute nodes.

```
BootStrap: docker
From: quay.io/centos/centos:stream8

%files
/opt/MLNX_OFED_LINUX-5.8-1.1.2.1-rhel8.7-x86_64.tgz
/opt/pmix-3.2.3.tar.gz
```

---

[10] https://www.izum.si/en/vega-en/.
[11] https://network.nvidia.com/products/infiniband-drivers/linux/mlnx_ofed/.
[12] https://github.com/forresti/osu-micro-benchmarks/tree/master.

```
/opt/libfabric-1.17.0.tar.bz2
/opt/openmpi-4.1.4.tar.bz2
/opt/osu-micro-benchmarks-7.0.1.tar.gz

%environment
  export OMPI_DIR=/usr/local
  export SINGULARITY_OMPI_DIR=$OMPI_DIR
  export SINGULARITYENV_APPEND_PATH=$OMPI_DIR/bin
  export SINGULAIRTYENV_APPEND_LD_LIBRARY_PATH=$OMPI_DIR/lib

%post

  ## Prerequisites
  dnf -y update
  dnf install --nogpgcheck -y dnf-plugins-core
  dnf config-manager --set-enabled powertools
  dnf groupinstall -y 'Development Tools'
  dnf install --nogpgcheck -y wget git bash hostname gcc gcc-gfortran \
    gcc-c++ make file autoconf automake libtool zlib-devel python3
  dnf install --nogpgcheck -y libmnl lsof numactl-libs ethtool tcl tk

  ## Packages required for OpenMPI and PMIx
  dnf install --nogpgcheck -y libnl3 libnl3-devel
  dnf install --nogpgcheck -y libevent libevent-devel
  dnf install --nogpgcheck -y munge munge-devel

  # Mellanox OFED matching VEGA
  mkdir -p /opt/mofed
  cd /opt/mofed
  tar -xf /opt/MLNX_OFED_LINUX-5.8-1.1.2.1-rhel8.7-x86_64.tgz
  cd MLNX_OFED_LINUX-5.8-1.1.2.1-rhel8.7-x86_64
  ./mlnxofedinstall --basic --user-space-only --without-fw-update \
    --distro rhel8.7 --force

  # PMIx
  mkdir -p /tmp/pmix
  cd /tmp/pmix
  tar -xf /opt/pmix-3.2.3.tar.gz
  cd pmix-3.2.3
  ./configure --prefix=/usr/local --with-munge=/usr &&\
    make -j
  make install

  # libfabric
  mkdir -p /tmp/libfabric
  cd /tmp/libfabric
  tar -xf /opt/libfabric*tar.bz2
  cd libfabric-1.17.0
  ./configure --prefix=/usr/local &&\
    make -j
  make install

  ## OpenMPI installation
  echo "Installing Open MPI"
  export OMPI_DIR=/usr/local
  export OMPI_VERSION=4.1.4
  mkdir -p /tmp/ompi
  cd /tmp/ompi
  tar -xf /opt/openmpi-${OMPI_VERSION}.tar.bz2

  # Compile and install
  cd /tmp/ompi/openmpi-$OMPI_VERSION
  ./configure --prefix=$OMPI_DIR --with-pmix=/usr/local --with-libevent=/usr \
    --with-ompi-pmix-rte --with-orte=no --disable-oshmem \
    --enable-mpirun-prefix-by-default --enable-shared --with-ofi=/usr/local \
    --without-verbs --with-hwloc
  make -j
  make install
```

```
# Set env variables so we can compile our applications
export PATH=$OMPI_DIR/bin:$PATH
export LD_LIBRARY_PATH=$OMPI_DIR/lib:$LD_LIBRARY_PATH
export MANPATH=$OMPI_DIR/share/man:$MANPATH

## Example MPI applications installation - OSU microbenchmarks
cd /root
tar -xf /opt/osu-micro-benchmarks-7.0.1.tar.gz
cd osu-micro-benchmarks-7.0.1/
echo ''Configuring and building OSU Micro-Benchmarks...''
./configure --prefix=/usr/local/osu CC=$(which mpicc) CXX=$(which mpicxx) \
  CFLAGS=-I$(pwd)/util
make -j
make install


%runscript
  echo ''Container will run: /usr/local/osu/libexec/osu-micro-benchmarks/mpi/$*''
  exec /usr/local/osu/libexec/osu-micro-benchmarks/mpi/$*
```

4   Automated Container Deployment and Security

Some larger projects and experiments, such as ATLAS, provide optimised container images for their users and make them available via read-only mounted remote filesystem CVMFS[13] or image registries. In contrast, for other projects, such as Belle II, as well as for various smaller projects, the images are provided and maintained by the SLING administrators and are shared with the SLING community, either by CVMFS or other means. However, for individual users and smaller projects that require specifically customised environments and have the expertise to create their own containers, it is logical to provide them with the ability to create custom containers for their workloads. In such cases, Vega, Arnes and NSC enable the Apptainer/Singularity --fakeroot feature, which emulates full privileges inside isolated user namespace i.e. with the appearance of running as root[14]. Upon request from the user, the login nodes are configured to include subordinate user IDs and group IDs, enabling the construction of Apptainer images with the --fakeroot feature. Building the container images directly on the production cluster where the workload is executed brings additional performance benefits, as the software can be optimised to the pertinent hardware and the interconnect.

Another important aspect in container deployment is security, given the sensitivity and value of the data and computations involved. Ensuring secure software development and deployment in HPC requires multiple layers of security checks [4], which can be included throughout the development lifecycle. To minimise the human errors in this process, automation is key to success. Security checks shall be implemented in the continuous integration/continuous delivery (CI/CD) process. This includes integrating static and dynamic code analysis tools to detect vulnerabilities and enforce coding standards before the software

---

[13] https://cernvm.cern.ch/fs/.

[14] https://apptainer.org/docs/user/main/fakeroot.html.

reaches the deployment stage and during the execution. Additionally, containerisation tools like Docker and Singularity can provide isolated environments that help secure the build and deployment processes by ensuring consistent and controlled software environments, using seccomp profiles, Linux Kernel Capabilities etc.

Implementing CI/CD is thus crucial in modern software development, streamlining the process from code changes to software release. CI involves frequent code commits to a central repository, automatically tested for potential issues. CD extends this, automating testing, building, and delivering software improvements for rapid and reliable releases. However, adapting CI/CD to HPC environments presents significant challenges. The complexity and diversity of HPC setups, along with the need for efficient resource utilisation and security management, multi-tenancy, secure access to resources and batch system interaction complicate automated testing and deployment. Specialised tools and approaches are necessary to integrate CI/CD into HPC effectively.

With SLING clusters, we're tackling the integration of CI/CD in HPC environments with a focus on security. Our approach is to integrate security checks into the GitLab CI/CD pipeline, using tools for static and dynamic application security testing, container image scanning and secrets management. To test code on the target infrastructure, we're exploring Docker-in-Docker (DinD) as a secure solution that ensures process isolation and reproducibility across different pipeline stages. After successful DevOps processes, software is made available through repositories such as CVMFS and building containers in registries such as Harbor, optimised for specific CPU and GPU architectures. A secure supply chain is just one piece of the puzzle. For secure usage of containers, it should be complemented by signing the images and verifying the signature before running the container. Additional security measures must be implemented on the system to ensure isolation and security of running the workload in the container, such as configuring seccomp profiles or using sandboxes within the containers.

5   Industry Use of Public HPC Resources

As the computational power of HPC resources continues to grow, industry users and SMEs are increasingly turning to these resources to meet their needs. Commercial users, in particular, seek to enhance the business value and competitiveness of their operations. Their workloads tend to be more standardised and predictable, necessitating the ability to reproduce and reuse their results. This is achieved through the use of checkpointing and the same software environments, among other techniques. Furthermore, they are concerned with the cost-effectiveness of their operations, as inefficiencies in their workloads result in financial losses. Commercial AI HPC workloads include simulations, but may also involve product design, engineering, and manufacturing. They have more specific and demanding requirements concerning isolation, security, and data lifecycle management (processing the data, saving the data, and destroying the data). The data, models, and code lifecycle are required. In order to ensure the reproducibility of software environments, containerisation is a highly suitable solution, as it enables the packaging and execution of applications within isolated environments that are independent of the underlying hardware and operating system. Another crucial aspect is the portability of software environments, not only across different HPC systems but also to other infrastructures, including grid, private clouds, Kubernetes clusters, and HPC. As per Cohen and al. [2], this is in contrast to the academic users, who often rely on legacy software, and industry users, who often require very recent libraries and applications, which are also challenging to install on HPC systems. Furthermore, the use of containers is also beneficial in this context. In order for HPC systems to become a viable alternative to public clouds, they must undergo significant modifications. While price is undoubtedly a crucial factor for commercial users, it is not the sole determining factor. The ability to programmatically access resources via APIs, as well as the ability to orchestrate, are also important. Most industry users use their own solutions for managing their workflows, which are often tied to the public cloud. In order to make the transition of using HPC resources easier for them, enabling the use of such solutions on HPC should be provided (Slurm REST API).

6   Conclusion

As described in this article, the evolution of software within the realm of HPC is increasingly gravitating towards containerisation as a solution to meet the growing demands for portability and reproducibility. Currently, we do not see an alternative that matches the capability of containers to provide a consistent and reproducible environment for HPC users. As industry adoption of HPC accelerates, the need for reproducibility, particularly with the use of specialized or proprietary software, becomes more pronounced. Containers offer a viable option for such software, including legacy applications. Container deployment automation, often integrated within CI/CD pipelines, not only streamlines the deployment process but also introduces additional layers of security control, contributing to a more secure software supply chain. Moreover, the performance aspect of containers is noteworthy. They are designed to incur no or minimal overhead, ensuring that jobs run in containers are executed with efficiency comparable to those on bare-metal systems. This seamless integration of container technology into HPC environments signifies a pivotal shift in software management, aligning with the complex and evolving needs of modern computational research and industry applications. In SLING, we aim to address the user requirements and adapt the development activities to their needs.

References

1. Alvarez, D., O'Cais, A., Geimer, M., Hoste, K.: Scientific software management in real life: deployment of EasyBuild on a large scale system. In: 2016 Third International Workshop on HPC User Support Tools (HUST), pp. 31–40. IEEE, Salt Lake City, UT, USA (2016). https://doi.org/10.1109/HUST.2016.009, http://ieeexplore. ieee.org/document/7830454/
2. Cohen, J., et al.: Simplifying the Development, Use and Sustainability of HPC Software (2013). http://arxiv.org/abs/1309.1101
3. Gamblin, T., et al.: The Spack package manager: bringing order to HPC software chaos. In: Proceedings of the International Conference for High Performance Computing, Networking, Storage and Analysis, pp. 1–12. ACM, Austin Texas (2015). https://doi.org/10.1145/2807591.2807623, https://dl.acm.org/doi/ 10.1145/2807591.2807623
4. Gantikow, H., Reich, C., Knahl, M., Clarke, N.: Providing security in container-based HPC runtime environments. In: Taufer, M., Mohr, B., Kunkel, J.M. (eds.) High Performance Computing, vol. 9945, pp. 685–695. Springer International Publishing, Cham (2016). https://doi.org/10.1007/978-3-319-46079-6_48, http://link. springer.com/10.1007/978-3-319-46079-6_48, series Title: Lecture Notes in Computer Science
5. Gerber, L.: Containerization for HPC in the Cloud: Docker vs Singularity, p. 35
6. Hoste, K., Timmerman, J., Georges, A., De Weirdt, S.: EasyBuild: building software with ease. In: 2012 SC Companion: High Performance Computing, Networking Storage and Analysis, pp. 572–582. IEEE, Salt Lake City, UT, USA (2012). https://doi.org/10.1109/SC.Companion.2012.81, https://ieeexplore. ieee.org/document/6495863/
7. Keller Tesser, R., Borin, E.: Containers in HPC: a survey. J. Supercomput. **79**(5), 5759–5827 (2023) https://doi.org/10.1007/s11227-022-04848-y, https:// link.springer.com/10.1007/s11227-022-04848-y
8. Kurtzer, G.M., Sochat, V., Bauer, M.W.: Singularity: scientific containers for mobility of compute. PLoS ONE **12**(5), e0177459 (2017) https://doi.org/10.1371/ journal.pone.0177459, https://dx.plos.org/10.1371/journal.pone.0177459
9. Liu, P., Guitart, J.: Performance characterization of containerization for HPC workloads on InfiniBand clusters: an empirical study. Clust. Comput. **25**(2), 847–868 (2022). https://doi.org/10.1007/s10586-021-03460-8, https://link.springer. com/10.1007/s10586-021-03460-8
10. Torrez, A., Priedhorsky, R., Randles, T.: HPC container runtime performance overhead: at first order, there is none (2019). https://api.semanticscholar.org/ CorpusID:245351594
11. Xavier, M.G., Neves, M.V., Rossi, F.D., Ferreto, T.C., Lange, T., De Rose, C.A.F.: Performance evaluation of container-based virtualization for high performance computing environments. In: 2013 21st Euromicro International Conference on Parallel, Distributed, and Network-Based Processing, pp. 233–240 (2013). https:// doi.org/10.1109/PDP.2013.41

Adapting Agricultural Virtual Environments in Game Engines to Improve HPC Accessibility

Dirk Norbert Baker[1,2]([✉]) [iD], Felix Bauer[3] [iD], Andrea Schnepf[3] [iD],
Hanno Scharr[4] [iD], Morris Riedel[1,2] [iD], Jens Henrik Göbbert[2] [iD],
and Ebba Hvannberg[1] [iD]

[1] School of Engineering and Natural Sciences, University of Iceland,
Reykjavik, Iceland
[2] Jülich Supercomputing Centre, Forschungszentrum Jülich GmbH, Jülich, Germany
d.baker@fz-juelich.de
[3] Institute of Bio- and Geosciences 3: Agrosphere, Forschungszentrum Jülich GmbH,
Jülich, Germany
[4] Institute for Advanced Simulation 8: Data Analytics and Machine Learning,
Forschungszentrum Jülich GmbH, Jülich, Germany

Abstract. E-infrastructures deliver basic supercomputing and storage capabilities but can benefit from innovative higher-level services that enable use-cases in critical domains, such as environmental and agricultural science. This work describes methods to distribute virtual scenes to the GPU nodes of a modular supercomputer for data generation. High information density virtual scenes, containing $> 100k$ geometries, typically cannot be rendered in real-time without techniques that change the information content, such as level-of-detail or culling approaches. Our work enables the concurrent and partitioned coupling to the image analysis in such a way that the data generation is dynamic and can be allocated to GPU nodes on demand, resulting in the possibility of moving through a continuous virtual scene rendered on multiple nodes. Within agricultural data analysis, the approach is especially impactful as virtual fields contain many individual geometries that coexist in one continuous system. Our work facilitates the generation of high-quality image data sets, which has the potential to solve the challenge of scarcity of well-annotated data in agricultural science. We use real-time communication standards to couple the data production with the image analysis training. We demonstrate how the use-case rendering impacts effective use of the compute nodes and furthermore develop techniques to distribute the workload to improve the data production.

Keywords: Visualization · Computing Services · FSPM · Computer Vision

A. Azab and T. Malkiewicz (Eds.): NeIC 2024, CCIS 2398, pp. 152–167, 2025.
https://doi.org/10.1007/978-3-031-86240-3_11

1   Introduction

Analyzing camera images to extract agricultural plant information is a major bottleneck and one of the most challenging tasks in plant science [1–3]. Synthetic data generation is a strong contender to combat the scarcity of high-quality annotated data [4,5], especially with the tools available through modern graphics engines [6,7]. A scalable plant image generation pipeline could improve critical tasks such as training of statistical or Deep Learning (DL) models [8,9]. However, for virtual agricultural data to be useful, it must span realistic scales and contain functional information, such as root water uptake or photosynthesis. In this work we improve the scalability of rendering virtual environments in graphics engines leveraging current e-infrastructure capabilities, thus enabling a data-scarce domain to make more effective use of High-Performance Computing (HPC) resources.

Synthetic data generation is a method to model the input and output data in conjunction to allow for larger-scale training data sampling. A virtual environment involves the time-variant rendering of a scene, including objects, lighting, and movement. We refer to the *simulation* of a virtual environment as the computation of processes such as water flow or photosynthesis, which depend on plant structure. Synthetic data production for plant image analysis can be based on plant simulation models, though it might not depend on it, or not use a simulated plant model at all. Synthetic data provides a rigorous assessment of error margins as well as a scalable pipeline [6], as the employment of virtual ground-truth provides a definite error value that has no hidden effects, such as human labeling error.

Synthetic data can be used to pre-train DL-models to increase robustness towards new data. Scalability and robustness are valuable for decision making in agriculture, where multi-step algorithms impact assessments and actions [2,10]. One advantage of synthetic data is that it can be used to provide a baseline for evaluation [11], while it also provides high-quality annotated data [7]. To accommodate the need for state-of-the-art rendering pipelines, we are using the Unreal Engine (UE) as rendering framework, coupling it with plant simulation models to create virtual environments. To enable the coupling and concurrent computation of simulation model, virtual world, as well as DL framework, we are employing Synavis [12]. We developed Synavis specifically for the coupling of virtual environments with DL and simulation models.

A key aspect of virtual field simulation is the use of Functional-Structural Plant Models (FSPMs), such as CPlantBox [13,14]. FSPMs are statistical descriptions of plant traits (phenotypes) [13] and are calibrated using measurements and statistical optimization [14]. Because FSPM outputs are diverse [5,12,15], the generated scene configuration provides more diverse image data. CPlantBox in particular is useful since it was stochastically evaluated [16] and has proven applicability in replicating field experiments [17]. The functional simulation of plant systems is complex and requires coupling between all connecting systems to form a fully capable simulation model [14].

Modern agricultural fields contain tens of thousands of plants, optimizing for high density and yield. A digitized version of realistic crop fields requires resource management as well as an adaptive pipeline to suit different use-cases that have conflicting requirements. To allow for increased scalability, we have developed techniques that facilitate the distribution of large-scale generation of virtual scenes of digital crop fields on to multiple GPU nodes of an HPC system. Our technique requires little user-based alteration to the virtual environment created in UE. The need for both scalable virtual field simulation as well as cohesive data generation is met through the distribution of FSPMs across graphics engine instances. We enable comprehensive workflows to generate synthetic data on HPC systems. This paper provides relevant insight into previous work in all supporting domains and techniques in Sect. 2, before describing specific HPC scenarios to support plant science domains in Sect. 3. We provide a description of our experiments in Sect. 3.4 before we describe the measurement results in Sect. 4 and discuss them in Sect. 5.

2   Related Work

In context of our use-cases, there are key aspects that are embedded in partly disjunct domains that need to work together to form a fully formed pipeline. This section highlights important work in all parts of the pipeline, starting with synthetic data generation in UE. A well-known framework for a data generation framework with UE was developed by Qiu et al. [7], called UnrealCV, which uses filters and image-based commands to provide the ground-truth for these filters, such as object contours or scene depth. It uses a direct Python binding of UE to allow for the expedient generation of data sets. Using UnrealCV, both Zhang et al. [18] and McCormac et al. [19] produced effective pipelines for RGB camera depth estimation. Especially Zhang et al. highlight the interplay between algorithm performance and UE scene generation, showcasing that the data generation needs to be dynamically adapted for validation purposes, which would need to have special accommodation from the e-infrastructure provider. There are a number of DL applications[1] that have a baseline evaluation or also data set generation through UE, particularly surrounding agent-environment interaction, like trajectory optimization by Roberts et al. [20]. The visualization of virtual scenes to train agents has seen an overall increase in use, especially in industry, as described by Nassif et al. [21]. Particularly, Bondi et al. [22] developed a separate UE-based approach to train unmanned aerial vehicles in a controlled setting.

From general synthetic data generation, we are now highlighting work that focuses on generating plant synthetic data. Certain algorithms, such as leaf segmentation, can be trained using data that is generated through image augmentation, as shown by Ward et al. [4], who generate top-view leaf data with semantic segmentation. However, there are classes of problems where image-based synthetic data is not sufficient. One challenge in plant science is the measurement of small-scale features. The estimation of poses of animals is object-centered

---

[1] Publications based on UnrealCV can be found here.

detection of small-scale features, and Mu et al. [23] developed an approach to use UE to generate the appropriate training data. The recent adaption of pose estimation to plant science by Berrigan et al. [24] shows that the transposition of certain methods to plant science depend on large-effort acquisitions of data sets. To combat this, synthetic data using plant models can be used, but these models need biological validation. Thus, Morandage et al. [17] show a use-case for simulation model evaluation - by providing a synthetic field example and analyzing how well, from a parameter estimation view, the FSPM CPlantBox can describe the field data and how accurate the estimations are. Lobet et al. [15] created a generation pipeline for virtual root system images, showing another use of simulation model-based synthetic data generation. A generalized modeling framework such as Helios [5] can encompass virtual fields that include plants, as well as their surface structure and reaction to light influx. While CPlantBox itself is a stochastic description of the plant structure, as introduced by Zhou et al. [13], more recent advances in the coupling of FSPMs for functional processes developed by Giraud et al. [14] illustrate the descriptive power of these systems.

The generation of synthetic data, and the above mentioned methods, do not make efficient use of cutting-edge HPC resources of e-infrastructures, even though they overlap domains that individually have seen innovation regarding scalability. One particular approach is data-parallelism, which is the partitioning of a data set across nodes. A visualization service-based example of data parallelism has been developed by Aunmüller et al. [25] in their framework *Vistle*. Data parallel approaches require rethinking some approaches to rendering, but will yield a better efficiency of using Graphics Processing Unit (GPU) nodes. Data parallelism is expedient for some visualization techniques, such as isosurface visualization, but more difficult to adapt to others. For example, Larsen et al. [26] showcase a raytracing approach that is compatible with the paradigm of data parallelism and have adapted this image rendering approach, which commonly requires the whole data set, to function in parallel. Moreland et al. showcase their design concepts for highly-threaded data-parallel visualization approaches for the Visualization Toolkit (VTK) [27].

From an e-infrastructur point-of-view, our challenge is accommodating the use of synthetic data generation methods using UE, which have been proven to increase robustness, on HPC systems to allow these techniques to scale with the increasing computational demand of DL frameworks.

3 Methods

Our aim is to improve the accessibility of HPC systems, providing plant scientists with methods that enhance their data augmentation. To this end, here we showcase two experiments (E1 & E2) focusing on the GPU performance when generating increasing number of plants (E1) and the ability to optimize field partitioning for visualization of large fields with HPC resources (E2).

3.1   Software and Data Generation

For our analysis of large-scale field generation, we are using an application that was built using UE on a user system. To couple UE with other data providers as well as with the training framework, we are using the Synavis framework to ease the setup of connecting the individual services. In Synavis, we can connect the FSPM CPlantBox, which outputs geometries of its plant simulation, to the virtual scene for rendering. Information that is generated by the simulation can be used as reference or label data, allowing the training of a variety of scenarios, especially the validation of DL-models that estimate plant traits. This data is being rendered within UE on-demand and the coupling is not synchronous, meaning that most changes to the virtual environment are just-in-time, which also applies to the image generation. The coupling used in our approach is based on real-time communication and no data is being written to disc. Most of the setup is done in user Python scripts, as the virtual scenes can be fully dynamic. However, it might be preferable to introduce pre-designed environments of publicly available project files, which is possible by including the Synavis plugin in an already existing UE application.

We are using a virtual field setup that uses stochastically parameterized plants [28]. These plants are discrete node-link descriptions that each have transition probabilities assigned to their structure. The leaf calibration as well as the geometrization methods for CPlantBox have been described by Helmrich et al. [12]. Plant surface geometry is inferred from centerline splines evaluated in fixed resolutions. Image rendering is largely dependent on what is feasible to render in one instance of UE. Essentially, the more plants can be rendered per instance, the larger field of view can be rendered, resulting in reduced need for stitching for a wider view. Simplification of the individual geometries is done by scaling the geometry resolution. Due to simulation coupling employed in our pipeline, there is a mix of base geometries being rendered in the scene along with geometries that are being treated with dynamic hierarchical culling (such as by UE-Nanite [29]). The visualization module for CPlantBox generates geometry buffers that can almost immediately be written into GPU memory. UE uses procedural meshes for this task, which have a very simple base layer of transformation and collision support and mesh sections for geometry buffers. In some cases, geometry buffers are filled separately using individual organs to allow for instance-based segmentation. This is especially important for leaf counting tasks, which are indicators for leaf development [4] and thus plant growth stage.

Measurements from UE use virtual textures that act as rendering targets for scene capture cameras. Direct scene rendering, processing image information such as object distance or velocity (i.e. pixel movement relative to the camera), or object property quantification, use this proxy to allow for immediate dynamic measurements. We use Synavis to handle measurement prompts, such that the controller is able to extract arbitrary measurements. This is especially important in cases where there is a mix of different data sources, such as image data from the renderer as well as plant data from the FSPM. HPC infrastructure allows the scalability of the DL-model training, but the other components of the workflow

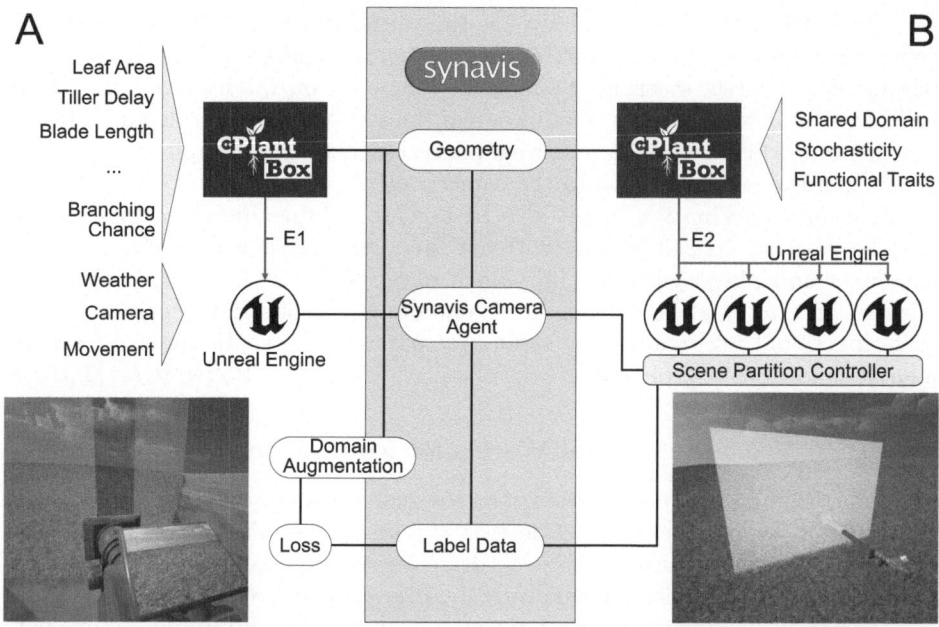

Fig. 1. Setup of pipelines in different distribution techniques. Measurements are highlighted and are also implemented in Synavis. The listed parameters are the primary steering parameters for the individual scenario. A. Continuous update with the intended use-case of adaptive data generation. B. Scene partition using Synavis and UE.

also need to be scalable to be on-par with the DL-model data requirements. To render the virtual scenes, UE requires the use of a Vulkan-compatible GPU, as is present in visualization or data analysis modules. Using Synavis, DL methods gain access to the domain-specific augmentation that is otherwise not accessible, which directly improves the pipelines that are also increasingly domain specific, with the potential for more impact.

3.2 HPC Scenario: Image Generation for Computer Vision

Computer Vision (CV) algorithms need labeled image data but often do not need to interact with the virtual environment. As such, compute infrastructure services for these systems should focus on enabling a responsive data generation to optimize DL training results. In our service implementation, Synavis keeps sending new FSPM realizations to UE, and parameter adaption directly influences the information content of rendered images. This is illustrated in Fig. 1.A, showing sample parameters that can be adapted in CPlantBox as well as in UE.

Ceaselessly updating the scene during image capture is a process that only functions if the relative speed of the camera agent compared to the simulation time is sufficiently slow. This limitation is largely dependent on how many plants need to be rendered, and how fast the FSPM computes a time step. In this setup, the DL framework typically has full authoritative control over the field generation, and all data as well as images being generated depend on initial parameters or direct steering. This method depends on the live coupling of the FSPM with the environment. Evaluation of the FSPM is done on-demand, but as the virtual world is fully dynamic, we input a ceaseless stream of plant geometries to place in the scene, relative to the camera agent.

Continuous evaluation might also be used to capture images of plant fields that have to be consistent but with no functional simulation, such as nutrient fluxes and photosynthesis [14], which have inherently competing elements. Absence of functional simulation, however, does not imply that the training data lacks functional information, as structural parameters can be fitted to experimental conditions like phosphorus availability, as done by Bauer et al. [28].

3.3   HPC Scenario: Virtual Worlds for Multi-agent Systems

Field partitioning becomes necessary at the scales that we see in agriculture, as rendering an average field size of 36.4 ha [30] in high detail requires distributed rendering. Infrastructurally, there needs to be an expedient pathway towards this partitioning that does not disturb the user-centered setup for these virtual environments, a challenge we meet through both explicit and implicit scene partitioning. We illustrate this approach in Fig. 1.B, which shows the setup with the partition controller that knows the partitioning boundaries and will connect the camera agent to a specific renderer when it enters its assigned area. The areas also define uniquely what plant structure is assigned to a specific location. An instance of the FSPM is assigned a seed at the start, and all organs are stochastic realizations of the input distributions of the parameter space [13]. The upscaling of the individual FSPMs to field level yields information on between-plant competition, for example to absorb sunlight. The field partitioning is stochastic seed based, which means that there would be structural (and thus functional) consistency between the individual compute nodes. The result is a fully informed virtual field that contains agricultural information, such as plant age, health, or leaf areas. This distribution is most effective if cameras are evenly distributed to nodes.

The partitioning, as seen in Fig. 1.B, is dependent on the preemptive assignment of regions to nodes. For a specific instance of UE, the simulation only generates a subset of the field, which in turn depends on how boundary conditions are being handles. For the purposes of the transition between rendering back-ends for a continuous camera path, we include a buffer region that is shared between neighboring nodes. This region is generated in addition and does not need to be communicated, as CPlantBox can generate identical structures on demand. Which rendering node is used is defined based on position on the field, which makes it necessary for the user to set the field partition in the run script or environment variables. Using Synavis for the scene partitioning allows the change of the camera source, depending on the position of the camera agent relative to the boundaries of the scene partitioning, as illustrated in Fig. 1.B. In our setups, we pre-register the streaming connection between endpoints on

Table 1. Node configuration for our tests. Instances of UE are run on dedicated nodes.

Module	JURECA-DC
CPU	2x AMD EPYC 7742, 2 × 64 Cores, 2.25 GHz
Memory	512 (16 × 32) GB DDR4, 3200 MHz
GPU	4 × NVIDIA A100 GPU, 4 × 40 GB HBM2e
Network	2 × InfiniBand HDR (NVIDIA Mellanox Connect-X6)

the infiniband [31] network, which is dedicated to HPC users. This means that the initialization phase is being skipped and the receiving application already allocated communication ports.

3.4 Experimental Setup

We tested two configurations, which are also highlighted in Fig. 1.A. These tests were performed on nodes of the JURECA-DC GPU module [32], with the node configuration shown in Table 1. We evaluated the rendering performance through the use of Synavis, which induces the transfer of a video stream. To measure the use-case of a continuous update that is applicable to CV as described in Sect. 3.2, we tested the rendering of N instances of the FSPM CPlantBox in Experiment 1 (E1). We measured the frame time as reported by UE through Synavis and the GPU utilization via the graphics vendor driver software. Details on the software and data setup are described in Sect. 3.1. We ran UE on 4000 × 3000 pixel resolution, with VP9 encoding on CPU. The partitioning of the virtual field into instances of UE has a specific worst case, which is concurrently running the instances of UE on one GPU node. We evaluated this in E2, which is highlighted in Fig. 1.B, using four concurrent instances of UE, running with a constant stream of new geometries similar to E1, but on the same node. Here, we evaluated the frame time performance for $4N$ instances of the simulation model. In this case, we tested the framework within one module, which means that setups that need to bridge via Ethernet will be slower than our setup using Infiniband. The framework has a baseline workload resulting from rendering an empty sunlit scene, and a minimum duration for the handling of commands.

4   Results

The measurement of E1, seen in Fig. 2 yielded no fixed maximum field size, but a decreased efficiency with increasing plant number. We measured a slightly superlinear increase in frame time for more complex scenes while the efficiency of using the GPU node decreased. Figure 2 shows the relative rendering performance depending on the amount of plants rendered in the scene, each with 28 d of growth time. Frame time average increased to 0.09 s which is roughly 10.7 frames per second at about 10k plant geometries. Notably, we observe a decreased GPU utilization, which is due to GPU memory exhaustion, resulting

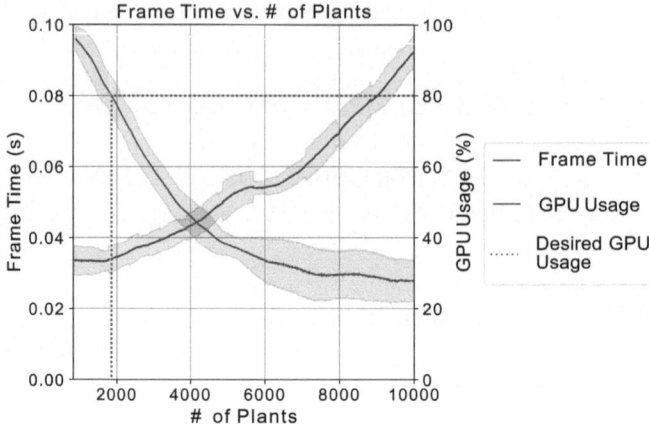

Fig. 2. Frame time measurement average for a field of N plants that are being continuously updated.

in more loading operations from the main memory. This results in a reduced efficiency of using the GPU node, which indirectly results in a decreased energy efficiency. While GPU usage is not the primary performance metric of the whole pipeline, we would like to keep this value above 80%, which referring to Fig. 2 yields an instance count of just below 2000 plants. As the simulation model and all new instances continue to be evaluated over time, Synavis constantly updates the scene and changes parameters. The highest impact to the rendering performance is entity creation, which refers to the spawning of a plant simulation model in the scene, along with registering of the object and handling of geometry information.

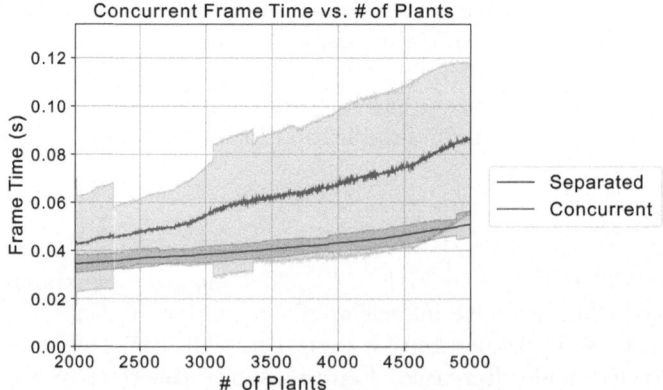

Fig. 3. Impact of rendering $N * 4$ geometries in UE concurrently on the same node as opposed to different nodes.

For E2, we show measurements of concurrent and non-concurrent rendering in UE on GPU nodes in Fig. 3. We observe, in the case of concurrently rendering four UE instances, a higher frame time on average as well as higher variance in frame time. The communication and geometric operations directly compete in this setting, for the Infiniband [31] network latency/throughput as well as in terms of memory operations. We note that we could not exclude effects from CPU usage competition between the instance of UE, because the base UE implementation does not respond to environmental variables and the parallel execution of UE threads might be suboptimal. The update time of individual plants in E1 is short enough that we are achieving a rendering time of more than 25 frames per second (0.04 s) for up to 3200 plants per GPU. An important note is that for distributed data generation, the image and environmental conditions such as weather and light need to be randomized to facilitate scalable training. Changing the settings of the virtual environment is typically done within 0.1 s from the time of prompting.

5   Discussion

The combination of simulation models, UE, and DL to enhance understanding and algorithms in plant science is a challenge both on a software level as well as infrastructural, as the necessary components have different requirements and best practices. Generally, synthetic data training is most effective with a foundational DL-model as basis, and subsequent real-world data fine-tuning. Synthetic data provides a scalable [5, 12] approach to generate a diverse [6, 18] data set but might not replicate all potential artifacts. In plant science, it is more impactful if rendered images contain biologically relevant information through plant simulation. Large field sizes that exceed GPU memory need to be partitioned once there are simultaneous observations in distant areas. This is partly because proper field upscaling should lift the plant model to the field scale at which agricultural decisions can be made. Our setup also allows multi-agent systems to communicate status information and the virtual scene can replicate visual information for each camera. The two main considerations that need to be discussed are how well our service-based infrastructural support performs with UE, and what we need to do to provide these pipelines to plant scientists.

5.1   Distribution Performance

The performance measurements were done purely from the standpoint of *rendering* performance. However, for a large-scale training use-case, the rendering performance might be secondary to an efficient use of tensor cores for the neural network training. Particularly if there need to be images with many plants in one view, users might forego the need for stitching and just assign all plants to one UE instance, resulting in a lower GPU utilization, which might not be critical in some cases. Furthermore, individual operations will reflect in the image data with a slight delay. Previous measurements in this framework have yielded

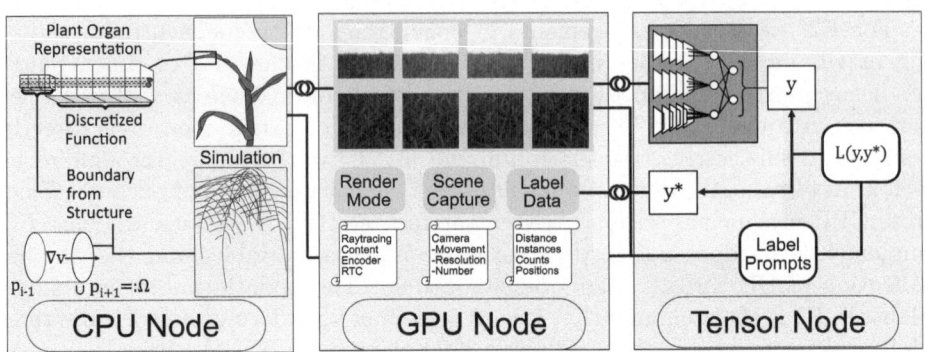

Fig. 4. Overview of technical components and data flows. The illustrated assignment to specific nodes is a performance recommendation, though individual components can share resources. A linked line indicates concurrent coupling.

a time difference between change request and confirmation of 0.1 s [12], but the execution of individual message workloads is done in bulk for certain operations (such as geometry data loading).

Streamlike data generation for CV is a suitable tool to create responsive and dynamic data that can change to the needs for the training framework. This is similar to active training frameworks, but the "rendering-in-the-loop" tuning of the virtual environment allows for a much more precise optimization of training impact. We achieve fairly good performance regarding the rendering of a large number of complex geometries. Our approach accommodates dynamic coupling such that we can partition large scales into individual instances through Synavis. As we are fully simulating the FSPMs, there are discrete updates to the scene geometry, which is not the case for pre-built geometries, but the approach is more scalable and directly visualizes biological information. The feedback loop between simulation and rendering allows for an exact quantification of either classification error thresholds (in domain-specific measures) or an analysis of robustness against scene conditions.

Dynamic camera coupling is a powerful tool to allow for a coherent field partitioning with multiple camera agents. We can render a much larger field by distributing the scene. Based on the desire to render efficiently with at least 80% GPU usage, we need to split an average of 50k plants per ha [33] into 25 GPUs per ha otherwise sacrificing GPU node efficiency as measured in E1 in Fig. 2. Transitions between individual nodes is reasonably fast with pre-registered connections (within 0.2 s). Drawbacks arise from the fact that there needs to be overlap between the areas of the field in cases of rendering simulation models to avoid information conflict, reducing the partitioning efficiency. Reconstructing a full video from this technique requires a few steps, e.g. a "fade" transition between streaming sources. Field partitioning allows for a better use of GPU resources, as we see that with more geometries on a single node (Fig. 2), the GPU utilization decreases, leading to less efficient use of compute time.

5.2   Infrastructural Support

Our example workflows consist of multiple components, that can be programmed together using Synavis. We acknowledge that the level of complexity involved in this system is high. However, due to the large amount of previous work in both the FSPM CPlantBox [13,14,16,17] as well as the compatibility provided by UE, new users will have an easier time adapting to each component individually. Due to the standardized method of communication between the frameworks, we are furthermore robust towards software version changes, which increases the inherent longevity of the framework in an everchanging HPC infrastructure landscape.

It is generally recommended that, for training purposes, multiple nodes are allocated, at least one of which would be a data generation node with UE. Figure 4 shows the components and what primary computing resource they utilize. Notably, the plant simulation produces *geometries* quite fast, while the *functional* coupling (particularly in the soil domain) requires fixed-point iteration solvers and thus might delay the digital plant evolution in cases of coupled growth. A functional coupling generally is most efficiently calculated on a shared memory system using all resources, while a purely structural simulation might run on the same node as UE. Our recommendation is to separate the neural network training and the data production, though it is possible to run these components on the same machine provided it has multiple GPUs. Particularly, this is one of the use-cases in which it is imperative to provide exclusive-use visualization or data analysis nodes to users. This is because shared GPU nodes which provide the only rendering capabilities in the system will inhibit certain scientific use-cases of HPC systems, especially in times of dedicated technologies such as tensor cores or massively-parallel GPUs as present in LUMI [34] or the planned JUPITER system [35].

6   Conclusion

Distributed rendering of plant fields for data generation is pertinent, especially in instances where large data sets, that would be cumbersome to store, are needed for the training. We enabled rendering virtual fields from plant simulations on nodes of the supercomputer JURECA-DC, highlighting continuous scene updates as well as field distribution as potential use-cases. With a strong basis of representative synthetic data, we have established techniques for distributed remote virtual environment rendering, and aim for large-scale use cases, such as light simulation [36], in the future. Furthermore, we want to extend our work on improving rendering performance and load balancing between nodes, depending on use-cases. We have shown that the parallel rendering and simulation of virtual environments is a valuable tool to establish a scalable data production pipeline and synthetic training environments for plant data analysis models, which is one of the most affected domains in terms of data scarcity and under-use of HPC infrastructure.

Acknowledgments. The authors would like to acknowledge funding provided by the German government to the Gauss Centre for Supercomputing via the InHPC-DE project (01-H17001).

This work has partly been funded by the EUROCC2 project funded by the European High-Performance Computing Joint Undertaking (JU) and EU/EEA states under grant agreement No 101101903.

This work has partly been funded by the German Research Foundation under Germany's Excellence Strategy, EXC-2070 - 390732324 - PhenoRob and by the German Federal Ministry of Education and Research (BMBF) in the framework of the funding initiative "Plant roots and soil ecosystems, significance of the rhizosphere for the bioeconomy" (Rhizo4Bio), subproject CROP (ref. FKZ 031B0909A).

The authors would like to acknowledge the compute time on the supercomputer JURECA [37] at Forschungszentrum Jülich GmbH, grant *VIS4AI*.

Data Availability Statement. Synavis is an open source repository and SynavisUE is its associated plugin for Unreal Engine. For new projects, we provide a template called SynavisUEexample. The CPlantBox official code can be found in the Plant-Root-Soil modeling group GitHub. For ensured compatibility and reproducibility, the branch associated with this paper has been forked separately. We have uploaded a video description of the method, seen in 10.6084/m9.figshare.25723773.

Disclosure of Interests. The authors have no competing interests to declare that are relevant to the content of this article.

References

1. Minervini, M., Scharr, H., Tsaftaris, S.A.: Image analysis: the new bottleneck in plant phenotyping. IEEE Signal Process. Mag. **32**(4), 126–131 (2015). https://doi.org/10.1109/MSP.2015.2405111
2. Taiz, L.: Agriculture, plant physiology, and human population growth: past, present, and future. Theor. Exp. Plant Physiol. **25**(3), 167–181 (2013). https://doi.org/10.1590/S2197-00252013000300001
3. Tsaftaris, S.A., Minervini, M., Scharr, H.: Machine learning for plant phenotyping needs image processing. Trends Plant Sci. **21**(12), 989–991 (2016). https://doi.org/10.1016/j.tplants.2016.10.002
4. Ward, D., Moghadam, P., Hudson, N.: Deep leaf segmentation using synthetic data. In: Proceedings of the British Machine Vision Conference (2018). https://doi.org/10.48550/arXiv.1807.10931
5. Bailey, B.N.: Helios: A scalable 3D plant and environmental biophysical modeling framework. Front. Plant Sci. **10** (2019). https://doi.org/10.3389/fpls.2019.01185
6. Zhang, T., et al.: UnrealPerson: an adaptive pipeline towards costless person re-identification (2020). arXiv:2012.04268
7. Qiu, W., et al.: UnreaLCV: virtual worlds for computer vision. In: Proceedings of the 25th ACM international conference on Multimedia, pp. 1221–1224 (2017). https://doi.org/10.48550/arXiv.1609.01326
8. Pound, M.P., et al.: Deep machine learning provides state-of-the-art performance in image-based plant phenotyping. GigaScience **6**(10) (2017). https://doi.org/10.1093/gigascience/gix083

9. Scharr, H., et al.: Leaf segmentation in plant phenotyping: a collation study. Mach. Vis. Appl. **27**(4), 585–606 (2016). https://doi.org/10.1007/s00138-015-0737-3
10. Kamilaris, A., Prenafeta-Boldú, F.X.: Deep learning in agriculture: a survey. Comput. Electron. Agric. **147**, 70–90 (2018). https://doi.org/10.1016/j.compag.2018.02.016
11. Mildenhall, B., et al.: Local light field fusion: practical view synthesis with prescriptive sampling guidelines. ACM Trans. Graph. (TOG) (2019). https://doi.org/10.1145/3306346.3322980
12. Helmrich, D.N., et al.: A scalable pipeline to create synthetic datasets from functional-structural plant models for deep learning. Silico Plants **6**(1), diad022 (2023). https://doi.org/10.1093/insilicoplants/diad022
13. Zhou, X.R., et al.: CPlantBox, a whole-plant modelling framework for the simulation of water- and carbon-related processes. Silico Plants **2**(1) (2020). https://doi.org/10.1093/insilicoplants/diaa001
14. Giraud, M., et al.: CPlantBox: a fully coupled modelling platform for the water and carbon fluxes in the soil-plant-atmosphere continuum. Silico Plants **5**(2) (2023). https://doi.org/10.1093/insilicoplants/diad009
15. Lobet, G., Koevoets, I.T., Noll, M., Meyer, P.E., Tocquin, P., Pagès, L., Périlleux, C.: Using a structural root system model to evaluate and improve the accuracy of root image analysis pipelines. Front. Plant Sci. **8** (2017). https://doi.org/10.3389/fpls.2017.00447
16. Schnepf, A., Huber, K., Landl, M., Meunier, F., Petrich, L., Schmidt, V.: Statistical characterization of the root system architecture model crootbox. Vadose Zone Journal **17**(1) (2018). https://doi.org/10.2136/vzj2017.12.0212
17. Morandage, S., Laloy, E., Schnepf, A., Vereecken, H., Vanderborght, J.: Bayesian inference of root architectural model parameters from synthetic field data. Plant Soil **467**(1), 67–89 (2021). https://doi.org/10.1007/s11104-021-05026-4
18. Zhang, Y., Qiu, W., Chen, Q., Hu, X., Yuille, A.: Unrealstereo: controlling hazardous factors to analyze stereo vision (2018)
19. McCormac, J., Handa, A., Leutenegger, S., Davison, A.J.: Scenenet RGB-D: can 5M synthetic images beat generic ImageNet pre-training on indoor segmentation? In: Proceedings of the IEEE International Conference on Computer Vision (ICCV) (2017). https://doi.org/10.1109/ICCV.2017.292
20. Roberts, M., et al.: Submodular trajectory optimization for aerial 3D scanning. CoRR abs/1705.00703 (2017)
21. Nassif, J., Tekli, J., Kamradt, M.: Digital Images – The Bread and Butter of Computer Vision, pp. 89–106. Springer Nature Switzerland, Cham (2024). https://doi.org/10.1007/978-3-031-47560-3_5
22. Bondi, E., et al.: AirSim-W: a simulation environment for wildlife conservation with UAVs. COMPASS 2018, Association for Computing Machinery, New York, NY, USA (2018). https://doi.org/10.1145/3209811.3209880
23. Mu, J., Qiu, W., Hager, G.D., Yuille, A.L.: Learning from synthetic animals. CoRR abs/1912.08265 (2019)
24. Berrigan, E.M., et al.: Fast and efficient root phenotyping via pose estimation. Plant Phenomics **6**, 0175 (2024). https://doi.org/10.34133/plantphenomics.0175
25. Aumüller, M.: The architecture of vistle, a scalable distributed visualization system. In: Markidis, S., Laure, E. (eds.) Solving Software Challenges for Exascale, pp. 141–147. Springer International Publishing, Cham (2015). https://doi.org/10.1007/978-3-319-15976-8_11

26. Larsen, M., Meredith, J.S., Navrátil, P.A., Childs, H.: Ray tracing within a data parallel framework. In: 2015 IEEE Pacific Visualization Symposium (PacificVis), pp. 279–286 (2015). https://doi.org/10.1109/PACIFICVIS.2015.7156388

27. Moreland, K., et al.: VTK-M: accelerating the visualization toolkit for massively threaded architectures. IEEE Comput. Graph. Appl. **36**(3), 48–58 (2016). https://doi.org/10.1109/MCG.2016.48

28. Bauer, F.M., et al.: In silico investigation on phosphorus efficiency of ZEA mays: an experimental whole plant model parametrization approach (2023). https://doi.org/10.34734/FZJ-2023-04031

29. Karis, B., Stubbe, R., Wihlidal, G.: A deep dive into unreal engine 5's nanite. In: SIGGRAPH (2021)

30. White, E.V., Roy, D.P.: A contemporary decennial examination of changing agricultural field sizes using landsat time series data. Geo: Geography Environ. **2**(1), 33–54 (2015). https://doi.org/10.1002/geo2.4

31. Pentakalos, O.I.: An introduction to the infiniband architecture. In: International CMG Conference (2002). https://doi.org/10.1109/9780470544839.ch42

32. Thörnig, P.: JURECA: Data centric and booster modules implementing the modular supercomputing architecture at Jülich supercomputing centre. J. Large-Scale Res. Facil. JLSRF (2021). https://doi.org/10.17815/jlsrf-7-182

33. Zhang, Y., Xu, Z., Li, J., Wang, R.: Optimum planting density improves resource use efficiency and yield stability of rainfed maize in semiarid climate. Front. Plant Sci. **12** (2021). https://doi.org/10.3389/fpls.2021.752606

34. Markomanolis, G.S., et al.: Evaluating GPU programming models for the Lumi supercomputer. In: Panda, D.K., Sullivan, M. (eds.) Supercomputing Frontiers. Springer International Publishing (2022). https://doi.org/10.1007/978-3-031-10419-0_6

35. Shapiro, A.: Nvidia grace hopper superchip powers Jupiter, defining a new class of supercomputers to propel AI for scientific discovery. NVIDIA Enterprise Networking Press Release (2023). https://nvidianews.nvidia.com/news/

36. Malenovský, Z., et al.: Discrete anisotropic radiative transfer modelling of solar-induced chlorophyll fluorescence: structural impacts in geometrically explicit vegetation canopies. Remote Sens. Environ. **263** (2021). https://doi.org/10.1016/j.rse.2021.112564

37. Krause, D., Thörnig, P.: JURECA: modular supercomputer at Jülich supercomputing centre. J. Large-Scale Res. Facil. JLSRF (2018). https://doi.org/10.17815/jlsrf-4-121-1

Enhancing Project Impacts Through Benefit Realization Management

Tomasz Malkiewicz[✉]

Nordic e-Infrastructure Collaboration (NeIC) and CSC – IT Center for Science Ltd., Espoo, Finland
tomasz.malkiewicz@csc.fi

Abstract. Having come about in the 1990s, benefit realization management is a relatively new, but proven concept for assessing and managing impacts of the projects. While introducing it into the existing organization structure may not be easy and does not come for free, the benefits yielding from having it in place outweighs the cost and pays off in a long term. The benefit realization management is discussed based on Nordic e-Infrastructure Collaboration, a global role model in its implementation across the entire organization.

Keywords: Benefit realization management · Impacts · Nordic Collaboration

1 Introduction

Benefit realisation management (BRM) is the work carried out to ensure that benefits are realized during and after the project execution phase, hence maximizing the projects' outcomes and impacts. BRM is a relatively new concept, having come about in the 1990s to address the frequent failures observed in the IT projects. In late 2000s, BRM concept has matured, solidifying BRM as a recognized discipline.

Nordic e-Infrastructure Collaboration (NeIC) has introduced BRM to its project portfolio in late 2010s. The BRM requires a strong foundation, i.e., mature project management. With experience from taking into use Tietoevery's Practical Project Steering (PPS) model in early 2010s, NeIC has implemented the BRM across the orgaznization i.e., entire project portfolio, ~15 projects. The key to succeed has been training and communication.

Recently, the BRM is evolving and adapting to the changing environment, in particular challenges and requirements driven by postpandemic effects and increased digitization. Similarly, at NeIC, the benefit realization is implemented in the fit-for-purpose fashion – in some projects there is a large emphasis on benefit realization management, whereas in the others the efforts are reduced to 'bare minimum', i.e., updating the benefit realisation management plans twice a year.

A. Azab and T. Malkiewicz (Eds.): NeIC 2024, CCIS 2398, pp. 168–177, 2025.
https://doi.org/10.1007/978-3-031-86240-3_12

2 Definitions and Motivation

2.1 Benefit Realization Management

A benefit is the measurable outcome of the project that is perceived to improve business value by the stakeholders. The benefit realization management (BRM) is a mean to support achieving the planned and unplanned project outcomes, i.e. benefits [1].

The main objectives of BRM are:

- increasing the likelihood of achieving the project outcome
- facilitating the organization to deliver success from its project investments
- improving organizational decision-making resulting in a more optimized project portfolio
- aligning better with organizations' strategy and encountering less resistance to change

The expected benefit consists of benefit objects, which make the benefit tangible and measurable. The business case is a document describing and clarifying the expected benefit and its value for the project partners, stakeholders and society.

2.2 Nordic Collaboration

Digital research infrastructure in the Nordic-Baltic region enables extensive collaborations on the topics perceived to have collective value as Nordic, i.e. societal structures and values (e.g. trust), and shared history (incl. Languages). A Nordic-Baltic collaboration on digital research infrastructure called NeIC, Nordic e-Infrastructure Collaboration [2], through its 20 development projects and activities, since 2012 has been developing services and sharing competencies for supercomputing, -data management, and training researchers and the next generation of software engineers.

NeIC's implementation of BRM is based on Tietoevery's Practical Project Steering, PPS [3], which is used for managing and steering NeIC's projects. For organizations like NeIC, BRM is a key tool to ensure that benefits are realized during and after the project execution phase, maximizing projects' outcomes and impacts.

2.3 NeIC Project Life Cycle

The lifecycle of NeIC projects consists of 4 phases: evaluation, preparation, execution and conclusion (highlighted with yellow in Fig. 1). In parallel to these project phases, the three benefit realization phases take place: evaluation, managing and realizing benefits (highlighted with red). After NeIC project ends, it enters the Affiliate programme (at the bottom right). The NeIC Affiliate Programme enhances the benefits and impacts of NeIC projects after they have been completed. All concluded NeIC projects and activities are eligible to participate, advertise their highlights on the Affiliate website and apply annually for NeIC financial support for the realization of benefit in post-project period.

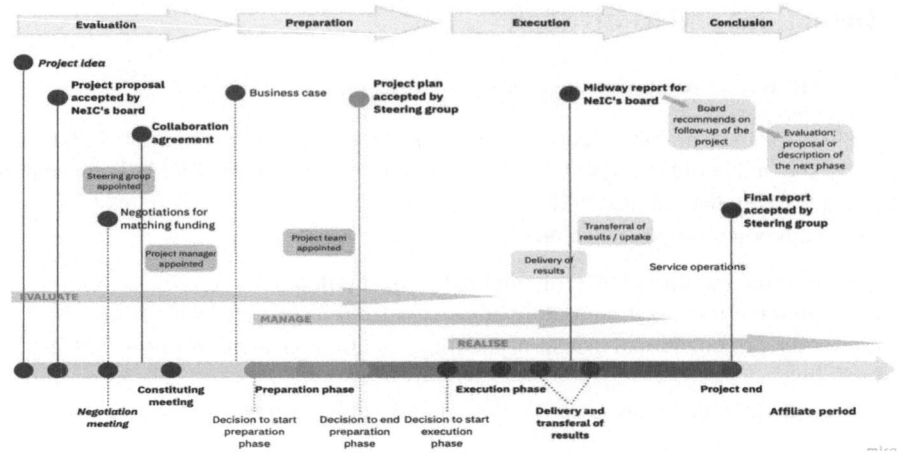

Fig. 1. Project lifecycle model implemented at NeIC.

3   Implementation of BRM

In early 2000s there has been a growing interest in BRM among professional bodies such as the Project Management Institute (PMI) [4], the Association for Project Management (APM), the Australian Institute of Project Management, and the International Project Management Association. Since then, the BRM field has seen its rapid growth, has been implemented in various forms by both public and private sector organizations and nowadays is recognized as a discipline of its own.

3.1   BRM at NeIC

As a part of the NeIC strategy to become a global role model in e-Infrastructure collaborations and develop further the organizational maturity level, NeIC adopted in 2014 the PPS model [3] as a project management tool. Further, in 2019 NeIC has developed a Policy of Benefit Realization Management (NeIC BRM) [5] and incorporated it into its' operations and the project portfolio. The policy, in addition to BRM model by PPS, has also been inspired by the PMI BRM framework by PMI. The purpose of the NeIC BRM is to support project portfolio management by forming a high-level foundation for decision-making regarding the project both by the founders and steering group of the project. Additionally, the policy provides precisely defined benefit realization and business change management framework for the projects. Similarly to PPS and PMI's BRM framework, the NeIC BRM divides benefit management into three stages: evaluating, managing, and realizing benefits. These three stages partially overlap, as depicted in Fig. 1.

Benefit Evaluation
Benefit evaluation begins soon after the project idea is brought forward. The project owner guides the stakeholders to determine the expected benefits and coordinates drafting a business case for the project based on these. Business case consists of the scope of the

project, list of relevant stakeholders, expected benefits, their valuation, risks and plan of realizing the benefits. Expected benefits consist of benefit objects and business changes needed to take place in order to achieve the benefits.

Benefit objects can be of money, time or quality type:

- monetary, e.g., increased revenues or reduced costs, valuated in money
- time, e.g., time savings can easily be converted into money, but the time actually saved must be utilised, in order to create benefit
- quality benefits are the most difficult to valuate. Instead of trying to assess the value of a quality benefit, it is given a value, based on how important it is compared to the money and time benefits. A well-composed benefit valuation group can, together, decide how important the quality benefits are in relation to the other benefits.

Division of the benefits into these three categories also helps deciding what business changes need to be done in order to provide the most value [1]. The monetary benefit object is valuated as money or equivalent, e.g. full-time equivalent (FTEs). The time and quality benefit objects are also often converted to FTEs. An example of benefit valuation is given in Sect. 3.3.

During benefit evaluation period the recipients of the project results and outcomes are identified and agreed upon. It is not an easy process, since organization are usually reluctant to take responsibility for the products and services that are not yet existing.

Time spent on the benefit valuation depends on the size of the project and partners' ambition. It is crucial that all relevant partners have identified the stakeholders, expected benefits, and business changes of the project.

Managing Benefits

The main tool supporting managing the benefits is the business benefit realization plan. It is created based on the business case once it has been approved by the project steering group. The benefit realization plan consists of benefit objects and business changes, an exemplary plan used by the NeIC iOBS project [6] is given in Table 1.

The business changes are organizational changes in operations and ways of working needed to take place in order achieve the benefits, example from NeIC iOBS is given in Table 2.

The business changes have no value in themselves but will deliver value through the benefits, upon completion. The benefits and business changes are therefore dependent on each other, i.e. benefit can be realized through business changes, which in turn are achieved by performing business change activities. Figure 2 presents the dependencies between the benefits, business changes and business change activities, according to the PPS model [3].

For large and complex projects, a separate owner for each benefit object and business change can be selected [1]. Otherwise, the project owner together with the steering group is in charge of monitoring the benefit objects and business changes. In the benefit realization plan it is documented who is in charge to initiate the business change, usually it is one of the steering group members from the partner organizations (see Tables 1 and 2). In case of risks of deviation from the planned business changes and benefits, the recovery plans are developed.

Table 1. Benefits in the benefit realization plan of the NeIC iOBS project.

Benefit realisation plan

Version 2.0, updated 2024-03-12, responsible: iOBS / PO

Benefit object	Measurable state, incl unit	Benefit realisation responsible	Benefit's importance for MetCoOp's achieving the	Project's contribution to the	Time	Plan and monitoring results										Summary	Updated

Tracked according to 80/20 rule

Improved Numerical Weather Prediction (NWP) forecast quality from increased number of observations used in data assimilation	Comparison of forecasts produced with and without the added weather observations (Q, $)	MetCoOp	4	5	Plan	0	0	0.5	1	1	1	1	1	1	2024-03-12: iOBS started the process of establishing the group looking into updates how MetCoOp is treating observations. Improving decoding and recoding process within SAPP (decodes, translate the observations (templates, data types) into SAPP routine and recodes from SAPP into the model). 2023-09-19: iOBS gave a kick-start for establishing the collaboration. Dig into SAPP in the	3/12/2024
					Result	0	0	0.5	1	1	1	1	1	1		
Improved quality control (QC) algorithms for pre-processing private observations where machine learning approaches might help to further identify important observation errors and/or instrument malfunctions	Comparison of forecasts produced with and without the added weather observations (Q, $)	MetCoOp	4	3	Plan		1	1	1	1	1	1	1	1	2024-03-12: Confident project ongoing, making software, based on the iOBS ideas. MetNorway is changing their quality control method, both private and public observations (feedback loop between the two, total amount for public+private of stations is about 10 000). 2023-09-19: Realised through Confident project (in the planning phase: 100 FTEs, now scaled down, 4 years)- list of software to be used for QC. iOBS WP2 software included in the development work.	3/12/2024
					Result		1	1	1	1	1	1	1	1		
Open access to an increasing volume of re-usable, quality-controlled data	Types of open access datasets (1. publicly funded weather observations, 2. comparison forecasts) (T, $)	MetCoOp, in cooperation with national meteorological institutions	3	4-5	Plan	1	1	2	2	2	2	2			2024-03-12: More and more FAIR data, available for the public. 2023-09-19: Publishing observations data in user-friendly way - people using what they've learned through iOBS. Now APIs are being taken into use, the result will be more open access. 2023-01-23: Publishing somewhat easier after the project ended, hard to keep track, people are building on iOBS work. What could be done within	3/12/2024
					Result	1	1		1.75	1.9	1.96	2	2			
Quality control for conventional observations. Starting point for collaboration within the Nordics.	Potential collaboration facilitated (S, Q) (1.) Collaboration established (2.)	National meteorological institutions	2	1	Plan			1	2	2	2	2	2	2	2024-03-12: Continued collaboration in the Nordics. MetCoOp (UWC East). Ideas from Confident project expected to be used by the other Nordic partners). 2023-09-19: Workflow being changed: quality control, storage and sending data. More tight together Nordic solution, using API and SAPP. UWC East in which Norway is part of	3/12/2024
					Result			2	2	2	2	2	2			

Table 2. Benefits in the benefit realization plan of the NeIC iOBS project.

Business change plan

Version 2.0, updated 2024-03-12, responsible: iOBS / PO

Business change	Description	Business change manager	Monitoring	Ready date	Summary	Deviation	Measures	Updated
Integration of new types of observations in national numerical weather predictions	Make changes to software so that private weather observations can be used as input to national weather forecast models in addition to the public observations traditionally used.	WP3 (MET Norway lead) and WP4 (SMHI lead)	MetCoOp, HIRLAM (name of the consortium responsible for the model at in 2020)	June 2021	2023-01-23: Confident project (run by MET NO) includes private observations data, other approach than initially envisaged. Other approach is iOBS code to MetCoOp for running on the model. Extended affiliate meetings continue bi-weekly (FI, NO, SE and EE) 2022-03-01: Not yet in operation, it's in testing. 2020: includes D4.4 2022-08-26: In testing at metCoOp, waiting for another software update.	Operations not part of Project plan.	Realised as planned: test in place.	3/12/2024
Improved quality control in numerical weather predictions	New observations types (private weather stations etc) provide data of unknown biases and metadata. They need a new kind of quality control to be useful in numerical weather prediction. Additionally, the increase in the total amount of observations data received by national meteorological institutes is suggesting a new approach to quality control using machine learning is needed.	WP3 (MET Norway lead) and WP2 (FMI lead)	MetCoOp	February 2020. June 2021	2023-01-23: Done at MET NO through Confident project and through yr.no (private observations used there for sometime, using probabilities to adjust the forecast). 2022-03-01: Shuffled into new project (Confident) 2020: includes report D3.2, software D2.2		Realised as planned: test in place.	3/12/2024
Future generations e-infrastructure design tool implementations and recommendations given	To accommodate future increased dataflows, we are testing new e-infrastructure chains using e.g. ECMWF's SAPP, machine learning software and dataflow tools to produce pipeline setup recommendations.	WP1 (MET Norway lead), WP2 (FMI lead) and WP5 (CSC lead)	MetCoOp	June 2021	2024-03-12: SMHI (SE) has new version of SAPP in place. 2023-09-19: New SAPP version, installed in NO. SE in the process of installing it. 2023-01-23: SAPP has been finally installed, close to be used as an operational dataflow, planned for Q2 2023. Object store, S3 (something like Google store but for computers, introduced by Amazon, installed at SMHI, data		Realised as planned: test in place at FMI.	3/12/2024

Realising Benefits

The third stage of the BRM, realizing benefits, begins when the project enters into the execution phase (see Fig. 1). The project owner together with the steering group monitors the progress of each benefit object and business change in the benefit realization plan, and regularly, e.g. quarterly, introduces necessary changes to the plan. Additionally, the project management (project management team and project owner) regularly informs the stakeholders about the benefit realization and business changes progress, projects' reference group, which meets usually 2–3 times per year, is a good channel for that. At the end of projects' execution phase, benefits and business changes are discussed at the projects' closing meeting/seminar as well as in the final report.

NeIC's 1-year Result transfer program, aims to support the transfer of the benefits to the recipient organizations and to monitor the impacts. The Result transfer program is offered mainly to the large projects that have essential unrealized benefits spanning beyond the project execution phase. The financing support is offered to follow result-transfer during the 1-year Result transfer period.

Once the project is finished, NeIC offers all the projects to enter into an Affiliate program, which is a continuous effort aiming at monitoring the impacts and benefits, and providing the projects with limited grants supporting benefit realization, e.g., to organize

Business benefit management

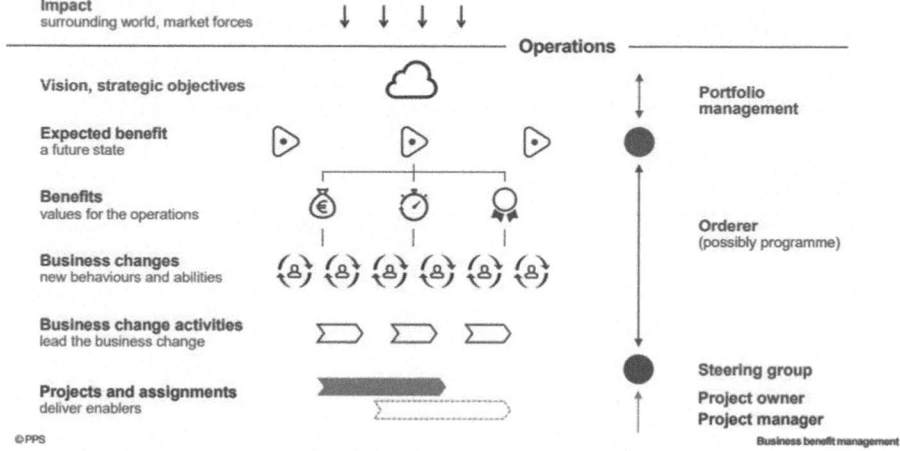

Fig. 2. Business benefit management based on PPS model [3].

thematic schools, partners and stakeholders' meetings or participate in conferences. All completed NeIC projects are currently members of the Affiliate program.

Many benefits are realized only after the project has ended. The temporary project organization, along with project resources and the steering group, is dissolved when the project is completed [1]. Therefore, it is crucial to have structure in place, such as 1-year Result transfer or Affiliate program, to monitor the benefits and support transferring of the results.

3.2 Examples of BRM Implementation in the Projects

While the project steering group and the project owner are mostly to define and decide upon the content of the business case, i.e., benefit objects and business changes, NeIC's Project Management Office (PMO) is in charge of implementing BRM within the projects, i.e., coordinate realization of the benefits. Tracking benefits during the projects' execution period is instrumental for its successful completion and assuring sustainability of the delivered results.

Valuation of the Benefits for the NMD Project

A benefit valuation allows comparing the benefit object's estimated profit to its anticipated cost. For instance, a cost could be the investment in project resources, continuous maintenance costs or costs of a change in operations [1]. The valuation is crucial to determine does it make sense to launch the project in the first place. At NeIC it is done in the project's business case, which is to be approved by the projects' steering group prior to the decision to start the project. Table 3 presents the summary of the benefit valuation for the Nordic Microdata Database project (NMD) [7].

Table 3. Summary of the benefit valuation for the Nordic Microdata Database project, numbers are given in k€, unless otherwise specified, [7] contains detailed descriptions.

NMD - estimate of value of database

Benefit object	Type of institution	# institutions	# researchers per institution	# projects per year	# uses per year	
			Total		1872	Per year

Project value	Years					
	1	2	3	4	5	6
Benefits realization	0 %	25 %	50 %	75 %	100 %	100 %
Values	0	468	936	1404	1872	1872
Costs (FTE)						
Total FTEs	2	2	2	2	0,5	0,5
Costs:	160	160	160	160	40	40
Annual net profit	-160	308	776	1244	1832	1832

net present value	3 713
Discount rate	10 %

In most cases it is impossible to come up with as accurate valuation of benefits as for the NMD. Even for the NMD, e.g. the economic value of more, better and more relevant research is not possible to estimate precisely.

Linking the BRM with the Project Plan

Once the business case is in place, the project plan is prepared by the project manager. In the project plan, the activities and deliverables with metrics are linked with benefit objects, example for the NeIC CodeRefinery3 project [8] is shown in Fig. 3. The deliverable is a tangible outcome, e.g., service or product, that needs to be produced with the selected activity in order to achieve the given benefit object. The deliverable is given a metric to monitor the progress and determine its completion [1].

A milestone is a set deadline when the deliverable needs to be completed. The project plan is presented and approved the project steering group, which provides high-level guidance: continuously prioritizes the benefit objects and sets deadlines for the deliverables and milestones. By steering group providing only this high-level guidance, one avoids micromanagement.

3.3 BRM Training

To achieve success with BRM, NeIC provides mandatory training for project managers, and also recommends it for all NeIC employees. NeIC offers two types of courses, level one and two, both playing a vital role in the organizational culture. Level one course covers the general aspects of NeIC's project management model, which includes BRM. Level two is tailored for project owners and project steering group members, with a focus on NeIC's BRM policy such as benefit evaluation and business change management. These trainings are conducted in cooperation with Tietoevry, strategic partner of NeIC since 2023. In addition, it is important to reflect on the lessons learned from the BRM after each project and apply them to the future ones.

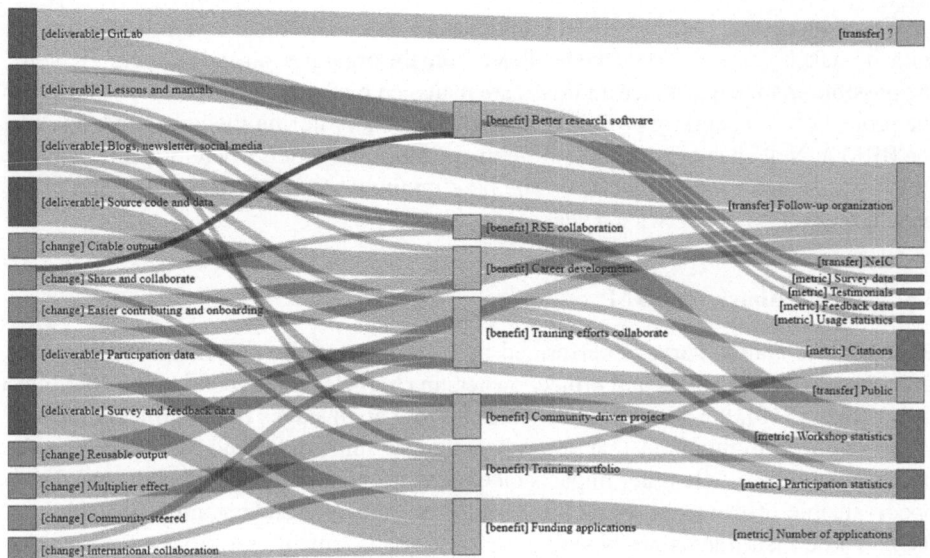

Fig. 3. Links between deliverables, business changes, benefits, metrics for measuring the benefits and recipients to which the outcomes are to be transferred in NeIC CodeRefinery3 project [8].

3.4 Change Management

BRM requires planning and people appointed to both steer and monitor the business changes and to steer the benefit realisation. The project orderer, i.e., NeIC on behalf of all the project partners, is officially responsible for this. However, in order to be successful, it requires commitment and engagement also from key stakeholders, especially the partner organisations in the project. Key terms are change management and resistance to change, the former being a systematic approach to dealing with the transition or transformation of an organization's goals, processes and technologies, and the latter unwillingness to adapt to new circumstances or ways of doing things. The purpose of change management is to implement strategies for effecting and controlling change and helping people to adapt to change.

Further, there is a need for the BRM buy-in both within the organization and among the stakeholders. For the motivating introducing the BRM into the organization, one can use the results of the PMI survey [9]: only 15% of the respondents rate their organization's project-related maturity level within the area of benefit and financial management as high. At the same time, there is a unanimous consensus of their importance: 93% of the respondents rate their importance as medium to high. It is further highlighted that the maturity level within the organization reflects the final project outcome. If the maturity level within the benefit and financial management area is definite high, the rate of project success is almost 4x higher compared to organizations with a low apparent maturity level.

Benefits are often challenging to realize, since organizational change is usually a complex process and takes time [1]. Thus, benefit realizations must be thoroughly planned and managed, and resistance to change needs to be factored in.

To be successful with the BRM implementation, one needs to continuously discuss with the stakeholders, evaluate the feedback from the organization and partners. At NeIC, the possible change resistance and risks are managed by the NeIC PMO, in particular by the project owners, and by project managers. Especially during the post pandemic era, the BRM at NeIC has evolved into more fit-for-purpose model, where projects dedicate to BRM as much time as practical. The bare minimum is reviewing and updating the benefit realization plans twice a year.

3.5 User Feedback on BRM

Based on the annual surveys performed by NeIC among the personnel and feedback from the project managers and project owners in charge of implementing the BRM into their projects [10], the BRM has been a useful tool at all stages of the project lifecycle.

It has been highlighted that some benefits are immediate and visible during the project's lifetime, while other impacts take many years to emerge, and NeIC has developed valuable instruments to track long-term impacts and understand how to support them. These instruments are benefit realisation management and project frameworks, and the Affiliate program.

The feedback loop is also beneficial for the organizations working on developing further project management models and BRM frameworks. NeIC strategically collaborates with Tietoevry, contributing to the development of the PPS model, providing feedback on the usefulness and the usage of PPS and its tools for managing NeIC project portfolio.

4 Conclusion

Benefit realization management is a proven tool for assessing and managing impacts of the projects. While introducing it into the existing organization structure may not be easy and will not come for free, the benefits yielding from having it in place exceed the cost and pay off on a long term.

NeIC showcases how one can introduce in a timely manner a novel way of managing both the projects and the project portfolio across the whole organization. The high-level goal of NeIC is to increase researchers' productivity and to enhance the quality of the research. To achieve this goal, NeIC included as a part of it's project life cycle a structured approach for benefits management, including benefits identification and realization. Implementing the benefit realisation throughout the NeIC project portfolio has been a major effort, especially change management, i.e., resistance to change. Based on the feedback from the stakeholders and project personnel, it has been worth it – both for the projects themselves (valuating and multiplying the outcomes) and for the organizations (NeIC partner and stakeholder organization implement alike models locally).

In order to be successful with BRM, organizations shall enhance employees' awareness of the significance of BRM and its implementation within the company by providing training as well as constantly monitor and improve the BRM practices within the organization.

References

1. Sandvik, A.: What are the Critical Success Factors for Benefit Realization Management? – Case Study, Master's thesis at Åbo Akademi University (2023)
2. NeIC - Nordic e-Infrastructure Collaboration, neic.no. Accessed 2024
3. Tietoevry. PPS Online. http://ppsonline.se/. Accessed 2024
4. PMI, Ahead of the Curve: Forging a Future-Focused Culture. Pulse of the Profession (2020)
5. Malkiewicz, T.: NeIC Policy for Benefits Realization Management. Nordic e-Infrastructure Collaboration (2019). https://wiki.neic.no/wiki/File:NeIC-Policy-for-BenefitsRealization-Management_approved-Aug-2019_updated-Oct-2019.pdf
6. NeIC iOBS project. https://neic.no/affiliate-iOBS/. Accessed 2023
7. Business case for the NeIC Nordic Microdata Database project (2024). https://wiki.neic.no/w/ext/img_auth.php/d/dc/Nordic_Microdata_Database_-_Business_Case.pdf
8. NeIC CodeRefinery3 project. https://neic.no/coderefinery/. Accessed 2024
9. PMI: Swedish Project Review - Decoding the path to digital adoption (2020)
10. Nordic e-Infrastructure Collaboration: How NeIC projects' benefits are realised (2023). https://neic.no/news/2023/04/21/Benefit-RealisationManagement/

Secure B2SHARE: Metadata Publication and Secure Sharing of Sensitive Datasets

Abdulrahman Azab[1]([✉]), Eirik Haatveit[2], and Harri Hirvonsalo[3]

[1] Sigma2 AS and University of Oslo, Oslo, Norway
abdulrahman.azab@sigma2.no
[2] USIT/University of Oslo, Oslo, Norway
eirik.haatveit@usit.uio.no
[3] CSC - IT Center for Science, Espoo, Finland
harri.hirvonsalo@csc.fi

Abstract. Secure B2SHARE is a framework that enhances the original B2SHARE service to allow for the secure sharing and management of sensitive datasets while ensuring compliance with privacy and data security regulations. This paper explores the concept of Secure B2SHARE, its architecture, the user journey through the system, and compares it with existing solutions for sensitive data management. The implementation of Secure B2SHARE at the TSD and CSC instances is discussed, showcasing how sensitive data is managed securely through these infrastructures.

1 Introduction

In recent years, the need for secure, compliant, and efficient ways to share and process sensitive data has become increasingly important. *Trusted Research Environments (TREs)* have emerged as a critical solution for managing sensitive data in a controlled and compliant manner. *TREs* allow researchers to process and analyze sensitive data within secure environments, ensuring that privacy and security requirements are upheld. Unlike traditional data archives, which are primarily designed for the long-term storage of data, *TREs* enable real-time data processing, where sensitive data remains within a controlled environment during analysis, preventing unauthorized access. In this context, *Secure B2SHARE* provides a robust framework for metadata publication and secure sharing of data that is being actively processed rather than archived. Extends the well-known B2SHARE service, which enables researchers to publish datasets in compliance with FAIR principles [2], *Secure B2SHARE* is tailored to support *secure metadata management* and *data sharing* within a *TRE*. It enables researchers to securely share metadata about sensitive datasets while ensuring that the data itself is not exposed to unauthorized access. By focusing on *active data processing*, *Secure B2SHARE* differs from *Sensitive Data Archives (SDAs)*, which are typically designed for storing data with limited active processing. While *SDAs* prioritize long-term preservation and access control for archived data, *Secure*

A. Azab and T. Malkiewicz (Eds.): NeIC 2024, CCIS 2398, pp. 178–182, 2025.
https://doi.org/10.1007/978-3-031-86240-3_13

B2SHARE enhances collaboration and ensures that data can be processed and shared in real-time while adhering to strict security standards.

This paper explores the *Secure B2SHARE* framework in greater detail, highlighting its architecture, its interaction with *TREs* like the *TSD (Tjenester for Sensitive Data)* and *CSC's Sensitive Data Services*, and the significant role it plays in the *European Open Science Cloud (EOSC)* ecosystem.

2   Secure B2SHARE Concept

Secure B2SHARE builds on B2SHARE by incorporating additional features specific to sensitive data management. These include secure data submission, metadata publication, and access control mechanisms [1]. The framework consists of three distinct components:

- **B2SHARE**: The base service that stores and makes metadata publicly discoverable [2].
- **Secure Data Submission Service (SDS)**: This component handles the secure upload of data into the system [7].
- **Authorization Service**: Controls access to datasets, ensuring only authorized users can access sensitive data [3].

Secure B2SHARE allows data owners to upload encrypted datasets through SDS and publish associated metadata in B2SHARE [4]. Access to sensitive data is only granted upon request and review by the data owner through the Authorization Service [5]. This ensures compliance with privacy regulations while maintaining the FAIR principles.

3   User Journey and Interaction Model

The user journey in Secure B2SHARE begins with the dataset owner uploading data to the SDS. Metadata is then described in B2SHARE, and the dataset is made publicly discoverable. When a researcher wishes to access a dataset, they search the B2SHARE interface or related metadata discovery services, such as B2FIND [6]. After finding a dataset of interest, the researcher submits an access request through the system. The request is reviewed by the data owner, who decides whether to grant or deny access. If access is granted, Secure B2SHARE notifies the secure storage system, allowing the researcher to access the data securely, with all data transfer happening within a secure environment. At the TSD instance, Secure B2SHARE relies on the REMS Resource Entitlement Management System for authorization, and at CSC, a similar setup is used with additional encryption processes for data storage [5] (Fig. 1).

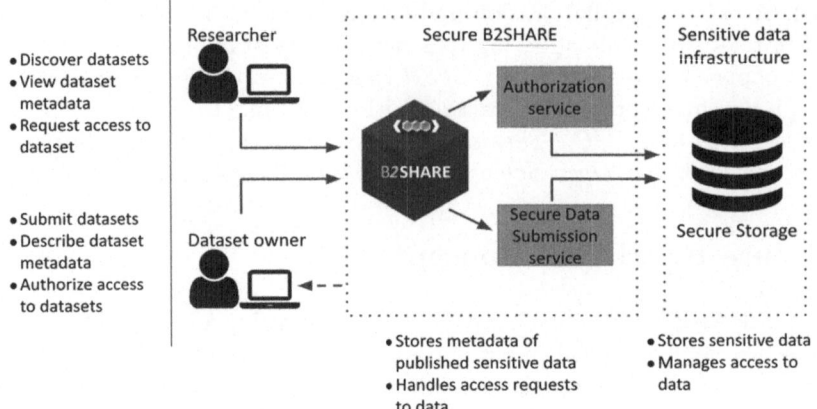

Fig. 1. Secure B2SHARE Interaction Model.

4   Secure B2SHARE Architecture at TSD and CSC

The implementation of Secure B2SHARE in sensitive data infrastructures such as TSD and CSC involves several key security measures. At CSC, datasets must be encrypted before submission and are stored in secure storage after re-encryption with local keys. Access to these datasets is provided through a secure remote desktop environment, where data is processed within the secure infrastructure, ensuring that sensitive data does not leave the secure environment [1].

The architecture at TSD is similar, with Secure B2SHARE managing the flow of metadata and access requests. The infrastructure at both sites ensures that data access is tightly controlled through identity federation, ensuring that the researcher requesting access is the same person authorized by the data owner [5] (Fig. 2).

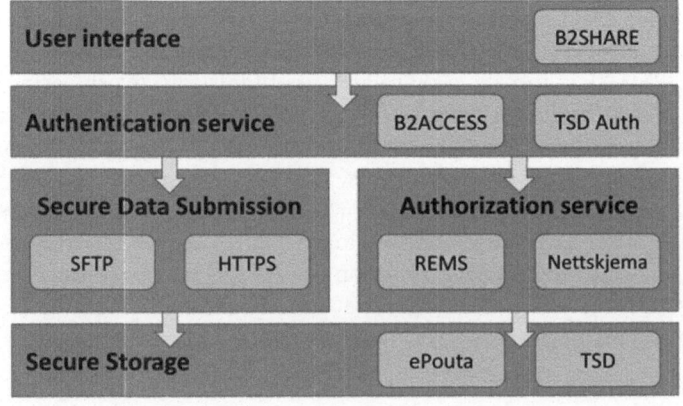

Fig. 2. Secure B2SHARE Layered Architecture.

5   Comparison with Other Solutions

There are several other solutions for secure sharing of sensitive data, such as Zenodo [7], Figshare [3], and the Federated EGA (European Genome-phenome Archive) [1]. These platforms allow metadata to be publicly accessible while restricting access to the data itself. However, Secure B2SHARE provides a more specialized and integrated approach, with tighter control over data submission, metadata management, and access authorization. Unlike other systems, Secure B2SHARE's interaction model ensures that sensitive data is never exposed without proper authorization, making it a more robust solution for sensitive data management. Additional platforms like the UK Biobank and health data repositories, such as the OpenHealthHub and the All of Us Research Program, offer similar features by enabling metadata discoverability and restricting access to the sensitive data upon review and authorization [8,9].

6   Conclusion and Future Work

Secure B2SHARE represents an important advancement in the management and sharing of sensitive datasets. The integration with existing sensitive data infrastructures at TSD and CSC ensures that data can be securely stored, shared, and processed without violating privacy regulations. Future work includes exploring the feasibility of a generic implementation of Secure B2SHARE across multiple infrastructures and improving the system's interoperability and usability for different research communities.

Acknowledgement. This work was supported by the DICE-EOSC project, funded by the European Commission under Horizon 2020 (Grant Agreement No. 101017207).

References

1. The European genome-phenome archive of human data consented for biomedical research. Nat. Genet. **47**(7), 692–695 (2015)
2. B2share: Data repository for the European open science cloud (2021). https://b2share.eudat.eu/. Accessed 11 Jan 2021
3. Figshare: Share your research (2021). https://figshare.com/. Accessed 11 Jan 2021
4. Openaire: Open access infrastructure for research in Europe (2021). https://www.openaire.eu/. Accessed 11 Jan 2021
5. Rems - resource entitlement management system (2021). https://www.csc.fi/en/research/rems. Accessed 11 Jan 2021
6. The TSD research platform for sensitive data (2021). https://www.uio.no/english/services/it/researchdata/tsd/. Accessed 11 Jan 2021
7. Zenodo - open access repository (2021). https://zenodo.org/. Accessed 11 Jan 2021
8. Jones, A., Smith, B.: The all of us research program: a model for inclusive health research. J. Health Inform. **25**, 123–130 (2021)
9. Smith, J., Wang, X.: The UK biobank: a resource for health data sharing. J. Publ. Health **44**, 789–795 (2020)

Enabling Message Passing Interface Containers on the LUMI Supercomputer

Alfio Lazzaro[✉][iD]

HPE HPC/AI EMEA Research Lab, Innere Margarethenstrasse 5, 4051 Basel,
Switzerland
alfio.lazzaro@hpe.com
https://www.hpe.com/emea_europe/en/compute/hpc/emea-research-lab.html

Abstract. Containers represent a convenient way of packing applications with dependencies for easy user-level installation and productivity. When running on supercomputers, it becomes crucial to optimize the containers to exploit the performance optimizations provided by the system vendors. In this paper, we discuss an approach we have developed for deploying containerized applications on the LUMI supercomputer, specifically for running applications based on Message Passing Interface (MPI) parallelization. We show how users can build and run containers and get the expected performance. The proposed MPI containers can be provided on LUMI so that users can use them as base images. Although we only refer to the LUMI supercomputer, similar concepts can be applied to the case of other supercomputers.

Keywords: Containers · Message Passing Interface · LUMI Supercomputer

1 Introduction

Containers represent a convenient way to package applications with dependencies for easy installation and productivity at the user level. They are used to package entire scientific workflows, software, libraries, and even data, solving the problem of making software run reliably when it is moved from one computing environment to another. They provide a simple way of sharing scientific applications and reproducing research on either cloud or High Performance Computing (HPC) systems. The *de-facto* standard technology is Docker, which is widely used in cloud environments [14]. However, mainly due to security concerns and the need to run a daemon on compute nodes, Docker containers are not suitable directly for running on HPC systems [7]. For this reason, optimized technologies have been developed for running on HPC systems such as Shifter [8], Charliecloud [17], Singularity [12], Sarus [4], Apptainer [6], or adapted for use on HPC, like Podman [10].

In this paper, we will discuss the procedures suggested for the deployment of HPC application containers on the LUMI supercomputer[1], specifically for

---

[1] https://www.lumi-supercomputer.eu/.

© The Author(s) 2025
A. Azab and T. Malkiewicz (Eds.): NeIC 2024, CCIS 2398, pp. 183–196, 2025.
https://doi.org/10.1007/978-3-031-86240-3_14

running applications based on Message Passing Interface (MPI) parallelization. LUMI provides the Singularity runtime for running containers. The Singularity *fakeroot* feature, which allows an unprivileged (*rootless*) user to build a container with the appearance of running as root [16], requires Linux user namespaces to be enabled. However, on LUMI user namespaces are disabled due to possible security risks concerns [2]. As a result of that, users cannot build their containers on LUMI, which is a desirable possibility for fast optimizing and testing of containers directly on the system where they will eventually run in production. To address these cases, Singularity (since SingularityCE 3.11) has introduced the possibility of building containers via `proot`, with some limitations[2]. We will describe how this technique has been deployed on LUMI in Sect. 2.

MPI is widely used by HPC applications to implement communications between parallel processes running on computed nodes connected by a high-speed interconnect. LUMI is an HPE Cray EX system that features the HPE Slingshot interconnect with 200 Gbps HPE Slingshot Cassini (CXI) network adapters [5]. The HPE Cray MPI is available on the system [11]. This is a proprietary and optimized implementation of the MPI specification, based on the open-source MPICH CH4 implementation from the Argonne National Laboratory [3]. HPE Cray MPI relies on the OFI Libfabric interface [9] with the CXI network provider to take advantage of the wide set of hardware capabilities offered by the HPE Slingshot network. Therefore, it is crucial to use the HPE Cray MPI to get high performance when running MPI applications in containers. This part will be described in Sect. 3, where we include some testing procedures and performance results.

2   Building Containers on LUMI

The Singularity version installed on LUMI is `singularity-ce version 3.11.4-1`. This version can build rootless containers when `proot` is available on the system `PATH`. On LUMI, this can be achieved via the module `systools`[3], provided by the LUMI User Support Team (LUST). An example of how to access the command is shown in Fig. 1. Then, Singularity containers can be built via the usual `singularity build` command.

In this paper, we are interested in building containers with MPI. Nevertheless, similar concepts can be considered for the building of any container. For running a Singularity container with MPI on LUMI we use the so-called *Hybrid* model[4] [19]. In this approach, we install a version of MPI inside the container to build it, but then we rely on the HPE Cray MPI available on LUMI during its execution. The prerequisite is that the MPI in the container must be compatible with the HPE Cray MPI. The HPE Cray MPI source code is not publicly available, however, HPE maintains the compatibility with the open-source MPICH

---

[2] https://docs.sylabs.io/guides/3.11/user-guide/build_a_container.html#unprivilged-proot-builds.

[3] https://lumi-supercomputer.github.io/LUMI-EasyBuild-docs/s/systools/.

[4] https://docs.sylabs.io/guides/3.11/user-guide/mpi.html#hybrid-model.

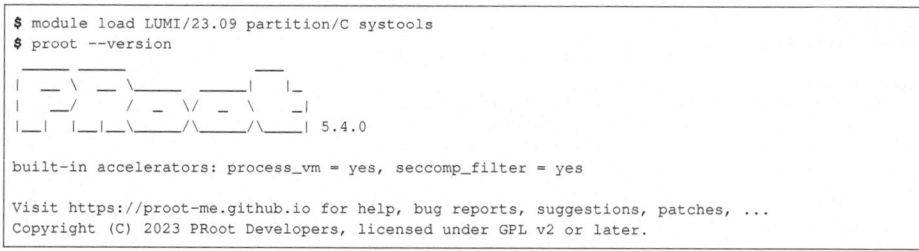

```
$ module load LUMI/23.09 partition/C systools
$ proot --version
   ____  ____
  |  _ \  _ \____  ____|  |_
  |  _/  /  _ \/  _ \   _|
  |__|  |__|_\____/\____/\___| 5.4.0

built-in accelerators: process_vm = yes, seccomp_filter = yes

Visit https://proot-me.github.io for help, bug reports, suggestions, patches, ...
Copyright (C) 2023 PRoot Developers, licensed under GPL v2 or later.
```

Fig. 1. An example of how to access the **proot** command (version 5.4.0) on LUMI via Lmod modules, provided by LUST. The command itself is provided by the **systools** module, which is in this example part of the software hierarchy provided by the modules **LUMI/23.09** and **partition/C**.

implementation, namely MPICH version 3.4a2 [11]. A minimal example of Singularity definition file for a container with MPICH 3.4a2 is shown in Fig. 2.

Although OpenMPI implementation [1] is not compatible with HPE Cray MPI, there is still a way to run a container based on it on LUMI. This is particularly useful since the OpenMPI implementation is quite popular as the default MPI implementation in most common Linux distributions. Our approach makes use of the HPE **MPIxlate** tool[5]. **MPIxlate** enables applications compiled using an MPI library that is not binary compatible with HPE Cray MPI, to be run without recompilation on supported HPE Cray systems. It does transparent runtime translation of the Application Binary Interface (ABI) between supported combinations of source and target MPI shared library implementations, in our case between OpenMPI version 4.x and HPE Cray MPI. A minimal example of Singularity definition file for a container with OpenMPI 4.1.6 (from the Ubuntu 24.04 release) is shown in Fig. 3.

A different approach for running OpenMPI containers on HPE Cray supercomputers has been proposed in Ref. [13]. The idea is to build the container with OpenMPI and OFI Libfabric support and then bind the LUMI OFI Libfabric library when executing the container, i.e. applying the hybrid mode at the level of the OFI Libfabric library. The drawbacks are that we don't make use of the optimized HPE Cray MPI and that we have to carefully install an OFI Libfabric library in the container compatible with the LUMI OFI Libfabric library, making the Singularity definition file more complex. For these reasons, we decided not to test this approach on LUMI, leaving its testing as future work.

So far, we have only discussed MPI implementations where the CPU buffers are used for communications. A separate discussion concerns the possibility of deploying containers with GPU-aware MPI functionalities, i.e. when the MPI implementation can send and receive GPU buffers directly, without having to first stage them in host memory. This is relevant for the LUMI GPU partition with AMD Mi250X GPUs. HPE Cray MPI supports GPU-aware MPI communications via the HPE proprietary GPU Transport Layer (GTL) library [11]. In

---

[5] https://cpe.ext.hpe.com/docs/mpt/mpixlate/index.html.

```
Bootstrap: docker
From: ubuntu:24.04

%post -c /bin/bash
  apt-get update && apt-get -y upgrade --no-install-recommends
  apt-get -y install --no-install-recommends \
          build-essential wget file ca-certificates \
          gfortran

  # Cleanup
  apt-get autoremove -y
  apt-get clean && rm -rf /var/lib/apt/lists/*

  # Installation dir
  export INSTALL_DIR=/container
  mkdir -p ${INSTALL_DIR}

  VER=3.4a2
  wget -q http://www.mpich.org/static/downloads/${VER}/mpich-${VER}.tar.gz
  tar xf mpich-${VER}.tar.gz && rm mpich-${VER}.tar.gz
  cd mpich-${VER}
  sed -i 's/libmpi_so_version="0:0:0"/libmpi_so_version="12:0:0"/g' configure
  FFLAGS='-fallow-argument-mismatch' \
    ./configure --prefix=${INSTALL_DIR}/mpi --disable-static \
                --disable-rpath --disable-wrapper-rpath \
                --enable-fast=all,O3 --with-device=ch3 \
                --mandir=/usr/share/man > /dev/null
  make -j$(getconf _NPROCESSORS_ONLN) install > /dev/null
  cd .. && rm -rf mpich-${VER}
  echo "export PATH=${INSTALL_DIR}/mpi/bin:\$PATH" >> ${SINGULARITY_ENVIRONMENT}
  echo "export LD_LIBRARY_PATH=\${LD_LIBRARY_PATH}:${INSTALL_DIR}/mpi/lib" >> \
    ${SINGULARITY_ENVIRONMENT}
```

Fig. 2. Minimal example of Singularity definition file to build MPICH compatible with the HPE Cray MPI available on LUMI. We use the Ubuntu 24.04 base image from Docker Hub. Note that we are specifically disabling the static libraries creation and the **rpath** of the shared libraries. Furthermore, we use the CH3 implementation, which doesn't require to install any network library interface, e.g. OFI Libfabric library. This allows binding the HPE Cray MPI when running the container.

particular, the GTL library has to be included during the linking phase of the applications within the containers, which will break the container portability. Furthermore, it is important to note that MPICH provides native GPU support only since version 4.x, which is not ABI compatible with HPE Cray MPI, and in general AMD GPU support is only available through the UCX communication framework, which is not compatible with the OFI Libfabric framework. Enabling OpenMPI with native GPU-aware on HPE Cray EX systems has been presented in Ref. [18]. Eventually, we will investigate how to extend our solutions to include GPU-aware MPI implementations as future work.

```
Bootstrap: docker
From: ubuntu:24.04

%post -c /bin/bash
  # fake some of the commands not available with proot
  for f in /usr/sbin/groupadd /usr/sbin/addgroup /bin/chgrp; do
      rm -rf $f
      ln -s /bin/true $f
  done

  apt-get update && apt-get -y upgrade --no-install-recommends
  apt-get -y install --no-install-recommends \
          build-essential wget libopenmpi-dev

  # Cleanup
  apt-get autoremove -y
  apt-get clean && rm -rf /var/lib/apt/lists/*
```

Fig. 3. Minimal example of Singularity definition file with OpenMPI (version 4.1.6), based on the Ubuntu 24.04 base image from Docker Hub. The **openssl** package, a dependency of the **libopenmpi-dev**, requires to run some privileged commands (**groupadd, addgroup, chgrp**), for which the **proot**'s emulation of the root user doesn't work. As a workaround, we *fake* them with symbolic links to the **true** command, so that they will always return without errors, allowing the successful build of the container. A more systematic approach to fake privileged commands has been proposed in Charliecloud via the kernel's **seccomp(2)** system call filtering [15].

3   Running Containers with MPI on LUMI

We use the MPICH container presented in Sect. 2 as a base image for building an application container. For our description, we consider a small MPI example (see Fig. 4) and the OSU Micro Benchmarks[6] as applications. In this way, we can test if MPI is properly running and its performance, respectively. An MPI application will crash with an error during the **MPI_Init** execution call if MPI is not properly initialized during the batch execution. On LUMI the workload manager is Slurm. However, it can also happen that the application doesn't report any error at the init phase, pretending to run multiple identical sequential instances with an eventual crash at a later stage of the execution. To avoid such a case, we adopt the good practice of including a test during the **MPI_Init** via a **LD_PRELOAD** intercept of the call to make sure that the requested number of MPI ranks in Slurm is effectively the same of the running MPI ranks (following the principle of *failing fast*). The source code of this test is shown in Fig. 5. We put everything in a Singularity definition file, presented in Fig. 6, which we build on LUMI.

---

[6] https://mvapich.cse.ohio-state.edu/benchmarks/.

```
#include "mpi.h"
#include <stdio.h>
#include <stdlib.h>

#define MPI_CALL(func, args)                                              \
  {                                                                      \
    int ierror = func args;                                             \
    if (ierror != MPI_SUCCESS) {                                        \
      printf("\nMPI error: %s failed (%s::%d)\n", #func, __FILE__, __LINE__); \
      exit(EXIT_FAILURE);                                              \
    }                                                                   \
  }

int main(int argc, char *argv[])
{
  int myrank = -1, nranks = -1, len = 0;
  char version[MPI_MAX_LIBRARY_VERSION_STRING];

  MPI_CALL(MPI_Init, (&argc,&argv));
  MPI_CALL(MPI_Comm_rank, (MPI_COMM_WORLD, &myrank));
  MPI_CALL(MPI_Comm_size, (MPI_COMM_WORLD, &nranks));
  if (myrank == 0) {
    printf("# ranks = %d (rank id = %d)\n\n", nranks, myrank);
    MPI_CALL(MPI_Get_library_version, (version, &len));
    printf("%s\n", version);
  }
  MPI_CALL(MPI_Finalize, ());

  return EXIT_SUCCESS;
}
```

Fig. 4. MPI test example (`mpitest.c`) used to check that MPI is properly running. A correct output reports the corrected number of MPI ranks and the version of the MPI implementation library.

The final step before running the container is to properly bind the host HPE Cray MPI into the Singularity container. This is achieved by setting the paths and libraries via the `SINGULARITY_BIND` and `SINGULARITYENV_LD_LIBRARY_PATH` host environment variables, receptively. We extract the values by compiling and running the `mpitest.c` (Fig. 4) directly on LUMI, i.e. without containers, under the command `strace -f -e trace=%file`[7]. Then, a script parses the output to extract the values of `SINGULARITY_BIND` and `SINGULARITYENV_LD_LIBRARY_PATH` variables. The HPE Cray MPI libraries (C and Fortran versions) paths are actually provided by the module `cray-mpich-abi`. For convenience, LUST provides the module `singularity-bindings`[8] that users can load to set the host environment.

We run some tests on LUMI to check that the container based on MPICH works as expected. In the following `srun` commands, we omit the account (`-A`) and the partition (`-p`) flags.

---

[7] https://man7.org/linux/man-pages/man1/strace.1.html.
[8] https://lumi-supercomputer.github.io/LUMI-EasyBuild-docs/s/singularity-bindings/.

```
#define _GNU_SOURCE

#include <dlfcn.h>
#include <stdio.h>
#include "mpi.h"
#include <stdlib.h>
#include <string.h>
#include <stdbool.h>

#define MPI_CALL(func, args)                                            \
  *ierror = func args;                                                  \
  if (*ierror != MPI_SUCCESS)                                           \
    return;

void mpi_init_check(int* ierror) {
  if (*ierror==MPI_SUCCESS) {
    const char* senv = getenv("CHECK_MPI");
    if (senv!=NULL && (senv[0]!='0' || strlen(senv)>1)) {
      bool senv_verbose = (senv[0]=='2' && strlen(senv)==1);
      int myrank, nranks;
      MPI_CALL(MPI_Comm_rank, (MPI_COMM_WORLD, &myrank));
      MPI_CALL(MPI_Comm_size, (MPI_COMM_WORLD, &nranks));
      const char* slurm_ntasks_env = getenv("SLURM_NTASKS");
      if (slurm_ntasks_env!=NULL) {
        const int slurm_ntasks = atoi(slurm_ntasks_env);
        if (slurm_ntasks!=nranks) {
          fprintf(stderr, "ERROR: # MPI ranks %d != # Slurm tasks %d (rank id = %d)\n",
                  nranks, slurm_ntasks, myrank);
          fflush(0);
          MPI_Abort(MPI_COMM_WORLD, EXIT_FAILURE);
        }
        else if (senv_verbose) {
          fprintf(stderr, "INFO: # MPI ranks %d == # Slurm tasks %d (rank id = %d)\n",
                  nranks, slurm_ntasks, myrank);
          fflush(0);
          MPI_CALL(MPI_Barrier, (MPI_COMM_WORLD));
        }
      }
      else {
        fprintf(stderr, "WARNING: SLURM_NTASKS variable not declared (rank id = %d / # ranks =
            %d)!\n", myrank, nranks);
        fflush(0);
        MPI_CALL(MPI_Barrier, (MPI_COMM_WORLD));
      }
    }
  }
}

// C intercept call
int MPI_Init(int *argc, char ***argv) {
  int (*original_MPI_Init)(int *argc, char ***argv);
  original_MPI_Init = dlsym(RTLD_NEXT, "MPI_Init");
  int return_value = (*original_MPI_Init)(argc, argv);
  mpi_init_check(&return_value);
  return return_value;
}

// Fortran intercept call
void mpi_init_(int *ierror) {
  void (*original_mpi_init_)(int *ierror);
  original_mpi_init_ = dlsym(RTLD_NEXT, "mpi_init_");
  (*original_mpi_init_)(ierror);
  mpi_init_check(ierror);
}
```

Fig. 5. MPI_Init intercept call (C and Fortran APIs) to be used via LD_PRELOAD. By default, the test is disabled. It can be enabled by setting the environment variable CHECK_MPI=1 (or 2 for a verbose mode). Then, the Slurm environment variable SLURM_NTASKS is compared to the number of running MPI ranks and the execution aborts if the two values differ. The test can be part of the base MPI container.

```
Bootstrap: localimage
From: mpich-ubuntu24.04.sif

%files
    mpitest.c /container/test_mpi/
    intercept.c /container/test_mpi/

%environment
    export LD_PRELOAD="/container/test_mpi/intercept.so"

%post -c /bin/bash
    apt-get update && apt-get -y upgrade --no-install-recommends

    # Cleanup
    apt-get autoremove -y
    apt-get clean && rm -rf /var/lib/apt/lists/*

    cd /container/test_mpi/

    # Compile mpitest
    mpicc -o mpitest.x mpitest.c

    # Compile the intercept library
    mpicc -shared -fPIC -ldl -o intercept.so intercept.c

    # Build OSU benchmarks
    OSU_NAME=osu-micro-benchmarks-7.4
    wget -q http://mvapich.cse.ohio-state.edu/download/mvapich/${OSU_NAME}.tar.gz
    tar xf ${OSU_NAME}.tar.gz
    cd ${OSU_NAME}
    ./configure --prefix=/container/osu CC=$(which mpicc) CXX=$(which mpicxx)
    make -j$(getconf _NPROCESSORS_ONLN) install
    cd ..
    rm -rf ${OSU_NAME} && rm ${OSU_NAME}.tar.gz
```

Fig. 6. Singularity definition file, based on MPICH base image, used to test if MPI within the container is properly running (via the `mpitest.x` and the `intercept.so`) and its performance (OSU Micro Benchmarks).

– Check which MPI library is used at runtime:

```
$ srun -n 1 singularity exec mpich-test.sif /container/test_mpi/mpitest.x
# ranks = 1 (rank id = 0)

MPI VERSION    : CRAY MPICH version 8.1.27.26 (ANL base 3.4a2)
MPI BUILD INFO : Fri Aug 11  0:25 2023 (git hash 55e934a)
```

HPE Cray MPI is correctly recognized.
– Report an error (by setting `CHECK_MPI=1`) if the module **singularity-bindings** is not loaded:

```
$ srun -n 2 singularity exec mpich-test.sif /container/test_mpi/mpitest.x
# ranks = 1 (rank id = 0)

MPICH Version:   3.4a2
MPICH Release date:      Wed Dec 18 14:46:35 CST 2019
MPICH ABI:       12:0:0
MPICH Device:    ch3:nemesis
MPICH configure:             --prefix=/container/mpi --disable-static --disable-rpath
    --disable-wrapper-rpath --enable-fast=all,O3 --with-device=ch3
    --mandir=/usr/share/man
MPICH CC:        gcc    -DNDEBUG -DNVALGRIND -O3
MPICH CXX:       g++    -DNDEBUG -DNVALGRIND -O3
MPICH F77:       gfortran -fallow-argument-mismatch  -O3
MPICH FC:        gfortran   -O3

# ranks = 1 (rank id = 0)

MPICH Version:   3.4a2
MPICH Release date:      Wed Dec 18 14:46:35 CST 2019
MPICH ABI:       12:0:0
MPICH Device:    ch3:nemesis
MPICH configure:             --prefix=/container/mpi --disable-static --disable-rpath
    --disable-wrapper-rpath --enable-fast=all,O3 --with-device=ch3
    --mandir=/usr/share/man
MPICH CC:        gcc    -DNDEBUG -DNVALGRIND -O3
MPICH CXX:       g++    -DNDEBUG -DNVALGRIND -O3
MPICH F77:       gfortran -fallow-argument-mismatch  -O3
MPICH FC:        gfortran   -O3

$ CHECK_MPI=1 srun -n 2 singularity exec mpich-test.sif /container/test_mpi/mpitest.x
ERROR: # MPI ranks 1 != # Slurm tasks 2 (rank id = 0)
application called MPI_Abort(MPI_COMM_WORLD, 1) - process 0
[unset]: write_line error; fd=-1 buf=:cmd=abort exitcode=1
:
system msg for write_line failure : Bad file descriptor
ERROR: # MPI ranks 1 != # Slurm tasks 2 (rank id = 0)
application called MPI_Abort(MPI_COMM_WORLD, 1) - process 0
[unset]: write_line error; fd=-1 buf=:cmd=abort exitcode=1
:
system msg for write_line failure : Bad file descriptor
srun: error: nid005347: tasks 0-1: Exited with exit code 1
srun: launch/slurm: _step_signal: Terminating StepId=7046274.0
```

Without CHECK_MPI=1 (by default CHECK_MPI=0), we see that two sequential instances of the application are executed (see the two corresponding outputs) and the container MPICH is used.

- Among the OSU performance benchmarks, we report here the results of the pt2pt bandwidth test, when running two ranks on different nodes:

```
$ CHECK_MPI=1 srun -n 2 --ntasks-per-node=1 singularity exec mpich-test.sif \
    /container/osu/libexec/osu-micro-benchmarks/mpi/pt2pt/osu_bw

# OSU MPI Bandwidth Test v7.4
# Datatype: MPI_CHAR.
# Size      Bandwidth (MB/s)
1                 2.01
2                 4.03
4                 8.05
8                16.13
16               32.20
32               64.39
64              128.81
128             257.57
256             490.31
512             976.65
1024           1949.08
2048           3890.51
4096           7763.86
8192          14551.28
16384         17642.47
32768         19151.75
65536         20953.49
131072        21800.11
262144        22206.10
524288        22370.86
1048576       22475.48
2097152       22527.21
4194304       22553.94
```

These values are in agreement with the corresponding results obtained on LUMI without the container. Same conclusion has been found for the other OSU benchmarks.

We can adapt the Singularity definition file reported in Fig. 4 to build a container based on the OpenMPI base image, by changing the corresponding base image in the From: section. Furthermore, since we run this container under mpixlate, we cannot define the LD_PRELOAD intercept library into the %environment section to avoid conflict with the tool. We replace this section with a %runscript section:

```
%runscript
    export
        LD_PRELOAD="${LD_PRELOAD}${LD_PRELOAD:+:}/container/test_mpi/intercept.so"

    if test $# -eq 0 || test -z "$@" ; then
        bash -norc
    else
        sh -c "$@"
    fi
```

Then, we load the modules (in this order): singularity-bindings, cray-mpich, and cray-mpixlate. We need to prepend the mpixlate libraries path to SINGULARITYENV_LD_LIBRARY_PATH. Finally, we can run some tests, similar to what we did for the MPICH container, where we specify mpixlate -s ompi.40 -t cmpich.12 in the Slurm launch command (see example below).

- Check which MPI library is used at runtime:

```
$ srun -n 1 mpixlate -s ompi.40 -t cmpich.12 singularity run openmpi-test.sif \
     /container/test_mpi/mpitest.x
# ranks = 1 (rank id = 0)

MPI VERSION    : CRAY MPICH version 8.1.27.26 (ANL base 3.4a2)
MPI BUILD INFO : Fri Aug 11  0:25 2023 (git hash 55e934a)
```

HPE Cray MPI is correctly recognized. We can check that OpenMPI is used inside the container by simply running:

```
$ singularity run openmpi-test.sif /container/test_mpi/mpitest.x
# ranks = 1 (rank id = 0)

Open MPI v4.1.6, package: Debian OpenMPI, ident: 4.1.6, repo rev: v4.1.6, Sep 30,
    2023
```

- We tested the OSU benchmarks, which are reporting the same performance obtained for the MPICH container, e.g.:

```
$ CHECK_MPI=2 srun -n 2 --ntasks-per-node=1 -A project_462000031 -p standard-g \
     mpixlate -s ompi.40 -t cmpich.12 singularity run openmpi-test.sif \
     /container/osu/libexec/osu-micro-benchmarks/mpi/pt2pt/osu_bw
INFO: # MPI ranks 2 == # Slurm tasks 2 (rank id = 0)
INFO: # MPI ranks 2 == # Slurm tasks 2 (rank id = 1)

# OSU MPI Bandwidth Test v7.4
# Datatype: MPI_CHAR.
# Size       Bandwidth (MB/s)
1                   1.97
2                   3.94
4                   7.86
8                  15.76
16                 31.58
32                 63.13
64                126.20
128               252.03
256               480.91
512               961.12
1024             1919.49
2048             3840.13
4096             7652.95
8192            14536.25
16384           17704.78
32768           19263.25
65536           21124.01
131072          21829.02
262144          22223.10
524288          22423.76
1048576         22514.62
2097152         22556.77
4194304         22579.80
```

4   Conclusion

In this paper, we have presented how users can deploy containerized applications on the LUMI supercomputer via the Slurm workload manager, specifically for running applications based on MPI parallelization. We have shown how users can build and run containers on LUMI based on the two main MPI implementations, namely MPICH and OpenMPI. We have presented some examples to show their functionalities and performance, including some tests to check that they properly run. The proposed MPI containers can be provided on LUMI so that users can use them as base images, either deriving from the existing container or re-building from scratch. Similar concepts can be applied to the case of other supercomputers.

As future work, we aim to explore: the OFI Libfabric injection proposed in Ref. [13], testing of other MPI implementations (e.g. MVAPICH), the OpenMPI solution proposed in Ref. [18], and the possibility to include GPU-aware MPI support.

Acknowledgments. The work presented in this article has been carried out as part of LUMI HPE Centre of Excellence activities. The author would like to thank the LUMI User Support Team members. A special thank to Tiziano Müller and Harvey Richardson of the HPE HPC/AI EMEA Research Lab for the useful discussions and comments.

Disclosure of Interests. The authors have no competing interests to declare that are relevant to the content of this article.

References

1. Open MPI: Open source high performance computing. https://www.open-mpi.org/
2. CVE-2022-0185 (2022). https://nvd.nist.gov/vuln/detail/CVE-2022-0185
3. Argonne national laboratory: MPICH: A high-performance, portable implementation of MPI. https://www.anl.gov/mcs/mpich-a-highperformance-portable-implementation-of-mpi
4. Benedicic, L., Cruz, F.A., Madonna, A., Mariotti, K.: Sarus: highly scalable docker containers for HPC systems. In: Weiland, M., Juckeland, G., Alam, S., Jagode, H. (eds.) ISC High Performance 2019. LNCS, vol. 11887, pp. 46–60. Springer, Cham (2019). https://doi.org/10.1007/978-3-030-34356-9_5
5. De Sensi, D., Di Girolamo, S., McMahon, K.H., Roweth, D., Hoefler, T.: An in-depth analysis of the slingshot interconnect. In: SC20: International Conference for High Performance Computing, Networking, Storage and Analysis, pp. 1–14 (2020). https://doi.org/10.1109/SC41405.2020.00039
6. Dykstra, D.: Apptainer without setuid (2022). https://www.osti.gov/biblio/1886029
7. Gantikow, H., Reich, C., Knahl, M., Clarke, N.: Providing security in container-based HPC runtime environments. In: Taufer, M., Mohr, B., Kunkel, J.M. (eds.) ISC High Performance 2016. LNCS, vol. 9945, pp. 685–695. Springer, Cham (2016). https://doi.org/10.1007/978-3-319-46079-6_48
8. Gerhardt, L., et al.: Shifter: Containers for HPC. J. Phys. Conf. Ser. **898**, 082021 (2017)

9. Grun, P., et al.: A brief introduction to the OpenFabrics interfaces - A new network API for maximizing high performance application efficiency. In: 2015 IEEE 23rd Annual Symposium on High-Performance Interconnects, pp. 34–39 (2015). https://doi.org/10.1109/HOTI.2015.19

10. Heon, M., et al.: Podman: a tool for managing OCI containers and pods (2021). https://doi.org/10.5281/zenodo.4735634

11. Kandalla, K., McMahon, K., Ravi, N., White, T., Kaplan, L., Pagel, M.: Designing the HPE cray message passing toolkit software stack for HPE cray EX supercomputers. In: CUG2023 Proceedings (2023). https://cug.org/proceedings/cug2023_proceedings/includes/files/pap144s2-file1.pdf

12. Kurtzer, G.M., Sochat, V., Bauer, M.W.: Singularity: scientific containers for mobility of compute. PLoS ONE **12**(5), e0177459 (2017)

13. Madonna, A., Aliaga, T.: Libfabric-based injection solutions for portable containerized MPI applications. In: 2022 IEEE/ACM 4th International Workshop on Containers and New Orchestration Paradigms for Isolated Environments in HPC (CANOPIE-HPC), pp. 45–56 (2022). https://doi.org/10.1109/CANOPIE-HPC56864.2022.00010

14. Merkel, D.: Docker: lightweight Linux containers for consistent development and deployment. Linux J. **2014**(239), 2 (2014)

15. Phinney, M.: New Root Emulation Mode for Charliecloud Using seccomp. In: 2023 IEEE/ACM International Workshop on Containers and New Orchestration Paradigms for Isolated Environments in HPC (CANOPIE-HPC) (2023)

16. Priedhorsky, R., Canon, R.S., Randles, T., Younge, A.J.: Minimizing privilege for building HPC containers. In: Proceedings of the International Conference for High Performance Computing, Networking, Storage and Analysis. SC 2021, Association for Computing Machinery, New York, NY, USA (2021). https://doi.org/10.1145/3458817.3476187

17. Priedhorsky, R., Randles, T.: Charliecloud: unprivileged containers for user-defined software stacks in HPC. In: Proceedings of the International Conference for High Performance Computing, Networking, Storage and Analysis, pp. 1–10 (2017)

18. Pritchard, H., Naughton, T., Shehata, A., Bernholdt, D.: Open MPI for HPE cray EX Systems. In: CUG2023 Proceedings (2023). https://cug.org/proceedings/cug2023_proceedings/includes/files/pap140s2-file1.pdf

19. Sande Veiga, V., et al.: Evaluation and benchmarking of singularity MPI containers on EU Research e-Infrastructure. In: 2019 IEEE/ACM International Workshop on Containers and New Orchestration Paradigms for Isolated Environments in HPC (CANOPIE-HPC), pp. 1–10 (2019). https://doi.org/10.1109/CANOPIE-HPC49598.2019.00006

Author Index

© The Editor(s) (if applicable) and The Author(s) 2025
A. Azab and T. Malkiewicz (Eds.): NeIC 2024, CCIS 2398, pp. 197–198, 2025.
https://doi.org/10.1007/978-3-031-86240-3